SpringerBriefs in Electrical and Computer Engineering

For further volumes:
http://www.springer.com/series/10059

Nicolas Chapados

Portfolio Choice Problems

An Introductory Survey of Single and Multiperiod Models

 Springer

Nicolas Chapados
Department of Computer Science and Operations Research
University of Montreal
PO Box 6128
Succ. Centre-Ville
Montréal (Québec) H3C 3J7
Canada
chapados@iro.umontreal.ca

ISSN 2191-8112   e-ISSN 2191-8120
ISBN 978-1-4614-0576-4   e-ISBN 978-1-4614-0577-1
DOI 10.1007/978-1-4614-0577-1
Springer New York Dordrecht Heidelberg London

Library of Congress Control Number: 2011932222

Printed on acid-free paper

Springer is part of Springer Science+Business Media (www.springer.com)

Preface

This book grew out of an introduction to portfolio choice problems presented in Chapados (2010). This problem has a long history: Markowitz's 1952 treatment of the subject now known as "Modern Portfolio Theory" is close to reaching the venerable age of sixty. In the intervening years, notions such as mean-variance efficiency have had enormous impact in the theory and practice of finance, not only on the "mundane" task of asset allocation but as models of the general trade-off between risk and return in financial markets, as well as portfolio performance measurement and attribution.

For newcomers to the field, it has been increasingly difficult to obtain a broad yet concise coverage of the subject. On the one hand, the practitioner-oriented literature focuses, by and large, on single-period models and the techniques[1] needed do fix the deficiencies in Markowitz's simple quadratic programming formulation. On the other hand, more academic treatments address the elegant generalization to the multiperiod case, but have been far less accessible. Moreover, the substantial body of research outside the field of financial economics has largely been scattered, with no work attempting to bring a unified treatment to the topic.

This book aims to fill this gap by offering a broad coverage of portfolio choice, containing both application-oriented and academic results, along with abundant pointers to the literature for further study. It tries to cut through many strands of the subject, presenting not only the classical results from financial economics but also approaches originating from information theory, machine learning and operations research.

As such, it should prove useful to students entering the field as well as practitioners looking for a broad coverage of the topic.

---

[1] Which some would respecfully dub "hacks".

Acknowledgements

This work could not have existed without the direct and indirect contributions of numerous individuals.

Above all, my deepest gratitude goes to my graduate advisor, mentor, and friend, Yoshua Bengio. His tireless support, advice and insight helped me wander on profitable avenues.

I would like to thank my colleagues for helping me understand many of the obstacles of putting in practice nice theoretical constructions: Aaron Courville, Éric-Paul Couture, Christian Dorion, Charles Dugas, Réjean Ducharme, Jean-François Gagné, Christian Hudon, Alexandre Le Bouthilier, Louis-Martin Rousseau, Xavier Saint-Mleux and Pascal Vincent.

I would also like to recognize the insightful suggestions of Éric-Paul Couture for some chapters. Thanks are due to Yoshua Bengio, Michel Gendreau, Doina Precup and Pascal Vincent for reading and commenting on an earlier version of this work. And my sincerest thanks go to Christian Dorion for a detailed reading of a late draft and suggesting a number of important corrections. All remaining errors are, of course, my own.

Montréal, Canada,   *Nicolas Chapados*
April 2011

Contents

List of Figures

Chapter 1
Introduction

> *If economists could manage to get themselves thought of as humble, competent people, on a level with dentists, that would be splendid.*
>
> — *John Maynard Keynes*

PORTFOLIO CHOICE is a central problem of economic agents. In plain words, it asks how one should "best" spread one's wealth across a number of different assets to maximize return and control risk. Of course, each asset is unique and offers its own outcome perspectives. These can be roughly summarized by an "expected return" and a "risk" aspect: the first one quantifies what would be the likely price appreciation of the asset or income arising from it over a given time period; the second measures how uncertain these payoffs to the investor can be.

It has long been understood that there is a fundamental trade-off between these two aspects, yielding a continuum of opportunities: at one end of the spectrum, short-term government bonds provide very small returns with absolute certainty.[1] At the other end, small-cap growth stocks, for instance, may promise staggering returns—but only if the company succeeds, for otherwise the investor may as well completely lose all her money.

Just as important is how individual risks combine at the portfolio level: what is the overall portfolio risk if assets are combined in specific ways? Real-world assets are not independent; some may zig as others zag. This is the fundamental idea behind the concept of *diversification*: the overall portfolio risk may be *less* than the sum of the risks of the individual assets that constitute it.

These ideas are summarized in Fig. 1.1, which illustrates two hypothetical individual assets, 'government bonds' and 'stocks', on the risk–return plane. These two assets are assumed to have well-defined risk and return characteristics. The figure also illustrates the notion of "efficient frontier", which traces out the risk and return characteristics of *portfolios* mixing the individual assets in specific proportions. The

---

[1] Assuming that the bond is denominated in the country's national currency.

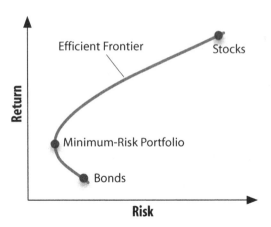

Fig. 1.1 Trade-off between risk and return in "modern portfolio theory". Portfolios on the efficient frontier yield the best return for a given risk level. Intermediate portfolios, mixing stocks and bonds, may exhibit lower risk than individual assets, a consequence of *diversification*.

key insight of diversification appears in plain sight: there exists portfolios whose risk is lower than either asset, but with better returns than the single lowest-risk asset.

The first quantitative treatment of diversification in portfolios of assets is due to the seminal paper of Markowitz (1952), who introduced, among other concepts, the notion of the efficient frontier on the risk–return plane; the methodology introduced by Markowitz, and perfected ever since by countless others, has been called Modern Portfolio Theory (MPT). However, an intuitive understanding of the benefits of diversification came much earlier. As Rubinstein (2002), in his half-century retrospective of Markowitz's paper, observes,

> Markowitz was hardly the first to consider the desirability of diversification. Daniel Bernoulli in his famous 1738 article about the St. Petersburg Paradox argues by example that risk-averse investors will want to diversify: "... it is advisable to divide goods which are exposed to some small danger into several portions rather than to risk them all together" (Bernoulli, 1738). As Markowitz (1999) himself points out in his historical review of portfolio theory, Bernoulli is also not the first to appreciate the benefits of diversification. For example, in *The Merchant of Venice*, Act I, Scene I, William Shakespeare has Antonio say:

> > "... I thank my fortune for it,
> > My ventures are not in one bottom trusted,
> > Nor to one place; nor is my whole estate
> > Upon the fortune of this present year ..."

> Although this turns out to be a mistaken security, Antonio rests easy at the beginning of the play because he is diversified across ships, places, and time.

Until Markowitz (1952), portfolio choice was approached on a "bottom-up" basis: each constituent (e.g. stock, bond) of the portfolio was chosen for its own risk and return characteristics, without regard for its interaction with the rest of the portfolio.[2] However, due to diversification effects, this simple form of analysis is in-

---

[2] Variance had been considered as a measure of financial risk as early as 1906 by Fisher (Fisher, 1906).

sufficient: the decision to hold a security should not only depend on a simple comparison of its expected risk and return profile to that of other securities, but also on its *marginal impact* on the risk–return profile of the investor's entire portfolio. Put differently, the decision to hold a security cannot be made in isolation, but is contingent upon the other securities that the investor already holds (or wants to hold). Earlier treatments of security analysis, including such classics as Graham and Dodd (1934) and Williams (1938), lack this perspective.

Myopia Dystopia

The original portfolio choice formulation by Markowitz has the investor make all her forecasts, of expected asset returns and covariances between them as we shall see in Chapter 2, at the start of an investment period, and then lets the investor rest until the end of the period. In particular, the investor is "prohibited" from tinkering with the allocations until the start of the next period. When that time comes, she acts as though any previous period never existed, or any further period will never exist: decisions are made strictly one period at a time. For this reason, Markowitz's formulation is called *single-period*.

It is also called "myopic", referring to the inability of the investor to see beyond the immediate future and anticipate future opportunities. Obviously, in practice, investors do not all die after one period, and a huge assortment of stratagems are employed to "repair" the single-period formulation to varying degrees and make it better reflect reality; Chapter 2 covers the most common ones.

However, even these fixes are insufficient since they do not reflect the fact that investment is, fundamentally, an extended process. The asset universe provides changing opportunities, some of which can be anticipated in advance. Perhaps the investor could want to consume a portion of her wealth along the way, or receives income from non-investment sources, changing the investable capital in known (or unknown) ways. Moreover, frictions abound in the process: there are costs to every trade, and governments are prompt to ask for a commission on any good deed (also known as "taxes"). Planning ahead for these contingencies, in fact for the complete future set of contingencies weighted by their probabilities, requires a drastically different viewpoint than that afforded by single-period approaches. They lead to the multiperiod formulations, first analyzed by Mossin (1968), Samuelson (1969) and Merton (1969) (see §3/p. 37).

A special group of investors commands specific requirements: that of *institutional investors*, in particular mutual or hedge fund managers operating in a competitive environment. Their main characteristic is that they are not only interested in maximizing the utility of their client's final wealth, but also in optimizing the trajectory that wealth takes to reach its final destination. Consider, for instance, a client choosing between two competitive funds offering similar returns; assuming that other fund characteristics are identical (including the stated investment risk profile), the client could well favor the fund having the "nicer" past return characteristics, where "nice" may not only include the variance of returns but more global criteria such as the *drawdown*.

Furthermore, on a day-to-day basis, the fund manager does not only care about how he will perform at the end of a long horizon, but how he is performing *right now*. It is a tired Wall Street cliché to state that "you are only as good as your last market call." Tired, perhaps, but a plea for help from practitioners that has seemed relatively ignored by academics.

Regrettably, the traditional multiperiod formulations outlined previously turn out to be unsatisfactory for the demands of institutional fund management. As stated, a fund manager operating in a competitive market cares as much about the path as about the final outcome.[3] In other words, the realized performance picture must be as rosy as possible, for as much of the time as possible, because clients can choose to join and leave the fund on a fairly unrestricted basis.[4] However, and this is a fatal mismatch, the utility functions assumed by the classical multiperiod solutions to the portfolio choice problem ignore these considerations and focus exclusively on the distribution of terminal wealth (generally in conjunction with an intermediate stream of consumption, which may be appropriate for a University endowment fund, but is irrelevant for a hedge fund or mutual fund manager). We argue that practitioners care about more dimensions of the picture than what has generally been assumed in the literature so far.

1.1 Overview

This work aims to review the main classical results about optimal portfolio construction, adopting a mostly-thematic rather than chronological perspective.

We start, in Chapter 2, with the classical single-period "modern portfolio theory" of Markowitz (1952; 1959) and its numerous refinements (§2/p. 7), including utility function variants, problem constraints, mean and covariance forecasting and econometric issues.

We then proceed, in Chapter 3, to the multiperiod and continuous-time formulations first studied by Mossin, Samuelson and Merton. A customary emphasis in this context has been to understand the structure of optimal solutions (under suitably analytically-tractable simplifications) and we shall examine the most enlightening of them.

Finally, straying from the traditional dynamic programming setting generally assumed in finance, we examine in Chapter 4 various "direct" and alternative criteria for portfolio choice, including stochastic programming and reinforcement learning, mainly studied in the machine learning and operations research communities.

---

[3] Other institutional investors, such as those working for defined-benefit pension funds, insurance companies, foundations and endowments are generally not subject to such stringent constraints.

[4] Although the financial panic of the Fall of 2008 has made long fund lock-up periods fashionable again, the trend until that point had been for lock-ups to become shorter in the competitive hedge fund industry, several funds offering redemptions with a 30-day notice or less.

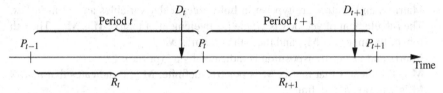

Fig. 1.2 Illustration of the time conventions followed in this document.

1.2 Basic Definitions and Notation

1.2.1 Simple Returns

In this work, we mostly consider the discrete-time scenario, in which *one period* (e.g. one day or one month) elapses between times t and $t + 1$, where $t \in \mathbb{N}$. We define period t to be the one elapsed between times $t - 1$ and t; see Figure 1.2.

Let $\{P_t\}, P_t \in \mathbb{R}_+$ be a random asset price process. We shall adopt the convention that any variable subscripted by a time index t can be measured given the set of information available at time t, which we denote $\mathscr{F}_t$.

Definition 1. The **simple rate of return** of an asset during period t is given by

$$R_t = \frac{P_t}{P_{t-1}} - 1.$$

For a dividend-paying asset, we consider dividends at time t, D_t, to be paid *immediately before* recording price P_t. The simple return taking dividends into account is

$$R_t = \frac{P_t + D_t}{P_{t-1}} - 1.$$

1.2.2 Risk-Free Asset

We denote by $R_{f,t}$ the rate of return earned by the risk-free asset (for instance, short-term government bonds).

1.2.3 Other Conventions

As much as possible, we attempt to adhere to the following notational conventions:

- Matrices and vectors are typeset in **bold face**; scalar variables are set in *italics*. The i-th element of vector $\mathbf{v}$ is $\mathbf{v}_i$; the i,j-th element of matrix $\mathbf{M}$ is $\mathbf{M}_{i,j}$. The i-th row of the matrix is $\mathbf{M}_{i,\cdot}$ and the j-th column is $\mathbf{M}_{\cdot,j}$.
- Matrix and vector transposition is indicated by a $'$ (prime).
- $\mathbf{M} \succ 0$ indicates that matrix $\mathbf{M}$ is positive-definite; $\mathbf{M} \succeq 0$ indicates that matrix $\mathbf{M}$ is semipositive-definite.
- It is sometimes useful to denote a vector of ones, whose size is appropriate given the context. We denote such a vector by the Greek letter iota, ι.

Chapter 2
Single-Period Problems

Risk is a part of God's game, alike for men and nations.
— *Warren Buffett*

I N THE SINGLE-PERIOD portfolio choice problem, the investor is assumed to make allocation decisions once and for all at the beginning of a given period (e.g. one quarter or one year), based on estimated prospects for the risk and return relationships of a universe of N investable assets over the horizon. Once made, the allocation decisions are not allowed to change until the end of the period; the impact of decisions arising in subsequent periods is not considered in this case, and for this reason, single-period problems lead to so-called *myopic* policies. Markowitz (1952) introduced the basic formulation, including expressions for the expected portfolio return and variance in terms of the portfolio weights and expected returns, variances and covariances of individual assets. He also introduced the *efficient frontier* and its depiction on the mean-variance plane. Since the original formulation uses the asset variances (and covariances) as the risk measure, the methodology is often called *mean–variance allocation*.

Despite their original conceptual simplicity, single-period problems are a large topic in which the optimization step is but one aspect. Just as important are the choice of utility function (§2.4/p. 11), risk measures (§2.5/p. 14), problem constraints (§2.6/p. 17) and forecasting models (§2.7/p. 23). Moreover, delicate issues related to the stability and econometrics of the obtained solutions need to be addressed for a successful implementation of the approach (§2.8/p. 29). This entails a rather involved methodology for single-period portfolio choice, which can be summarized by Fig. 2.1.

2.1 Basic Formulation

Let $\mathbf{R}_{t+1} \in \mathbb{R}^N$ be a vector of random *asset returns* between times t and $t+1$ (see §1.2/p. 5 for a summary of the time index conventions). Assume that the investor

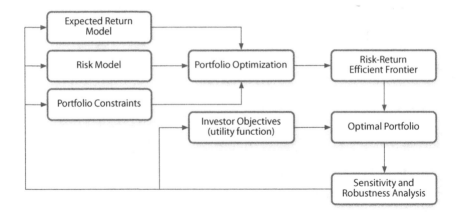

Fig. 2.1 Methodological steps surrounding the Markowitz single-period investment process; adapted from Exhibit 2.2 (p. 21) of Fabozzi *et al.* (2006).

makes, given the information available at time t, a forecast of the first two moments of the distribution of future returns,

$$\mu_{t+1|t} = \mathbb{E}_t\left[\mathbf{R}_{t+1}\right]$$
$$\Sigma_{t+1|t} = \text{Cov}_t\left[\mathbf{R}_{t+1}\right],$$

where the $\mathbb{E}_t[\cdot]$ and $\text{Cov}_t[\cdot]$ denote, respectively, the expectation and covariance matrix of a (vector) random variable conditioned on the information available at time t. For simplicity in this section, since single-period modeling does not explicitly consider the consequences of time, we drop the time subscripts on the above quantities, which we write simply as $\mathbf{R}$, μ and Σ. Likewise, the return on the risk-free asset during the period is denoted by R_f.

The investor allocates its capital among the N assets, forming a portfolio $\mathbf{w} \in \mathbb{R}^N$ where each element $\mathbf{w}_i$, the *weight* of asset i, represents the fraction of total capital held in the asset. The expected portfolio return and variance are given respectively by

$$\mu_P = \mathbf{w}'\mu \qquad \text{and} \qquad \sigma_P^2 = \mathbf{w}'\Sigma\,\mathbf{w}. \qquad (2.1)$$

We shall make the following assumptions about the assets:

1. There are no "redundant" assets, i.e. no asset return can be obtained as a linear combination of the returns of other assets.
2. All assets are risky (have positive return variance), which implies, in conjunction with the above assumption, that the covariance matrix Σ is nonsingular. (The inclusion of a risk-free asset is treated in §2.4/p. 11.)

Definition 2 (Efficiency). A portfolio **w** is said to be **efficient** if it is the lowest-variance portfolio for a given level of expected return.

The portfolio choice problem seeks to directly find efficient portfolios by determining an "optimal" vector of asset weights. The minimum-variance formulation of the problem considers the expected portfolio variance as the measure of risk. It takes the form

$$\mathbf{w}^* = \arg\min_{\mathbf{w}} \frac{1}{2} \mathbf{w}' \Sigma \mathbf{w} \tag{2.2}$$

$$\text{subject to} \quad \mathbf{w}'\mu = \rho, \tag{2.3}$$

$$\mathbf{w}'\iota = 1. \tag{2.4}$$

The objective function, eq. (2.2), seeks the vector of weights which minimizes the total expected portfolio variance, subject to constraint (2.3) which requires a portfolio return of ρ (which can be viewed as the desired or target return), and constraint (2.4) which specifies that all capital must be invested. We consider other types of constraints — and their implication on the solution methods — in §2.6/p. 17.

2.2 Solution

Since all constraints are of equality type, problem (2.2) can be solved analytically by introducing Lagrange multipliers. The general solution is derived in §A.1/p. 71. To borrow notation from that section, we set

$$\mathbf{A} = \begin{pmatrix} \mu' \\ \iota' \end{pmatrix} \qquad\qquad \mathbf{b} = \begin{pmatrix} \rho \\ 1 \end{pmatrix},$$

and obtain the optimal weights $\mathbf{w}^*$ by substitution into eq. (A.10). Some algebraic manipulation yields the somewhat simplified but enlightening form, showing the optimal weights $\mathbf{w}^*$ as being linear in the desired return ρ, (Merton, 1972; Fabozzi et al., 2007)

$$\mathbf{w}^* = \mathbf{g} + \mathbf{h}\rho, \tag{2.5}$$

where

$$\mathbf{g} = \frac{\Sigma^{-1}(c\iota - b\mu)}{d}, \qquad\qquad \mathbf{h} = \frac{\Sigma^{-1}(a\mu - b\iota)}{d},$$

and

$$a = \iota'\Sigma^{-1}\iota, \qquad b = \iota'\Sigma^{-1}\mu, \qquad c = \mu'\Sigma^{-1}\mu, \qquad d = ac - b^2.$$

Similarly, the globally minimum-variance portfolio (GMV) is obtained without imposing the expected-return constraint, yielding portfolio weights and variance respectively given by

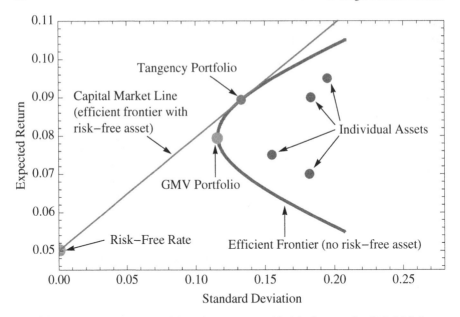

Fig. 2.2 Efficient frontier obtained from four assets specified in the text; the Global Minimum Variance (GMV) portfolio has a lower risk (as measured by the standard deviation of returns) than any individual asset, showing the benefits of diversification.

$$\mathbf{w}^*_{GMV} = \frac{\Sigma^{-1}\iota}{\iota'\Sigma^{-1}\iota} \qquad \text{and} \qquad \sigma^2_{GMV} = \frac{1}{\iota'\Sigma^{-1}\iota}. \qquad (2.6)$$

The above solutions yield two important insights. First, as will be illustrated next, it reflects the benefits of diversification. Second, it highlights that ultimately, higher returns can only be obtained by taking on higher leverage — thence more risk — since the optimal weight vector is linear in the target return ρ.

To illustrate these solutions, consider a four-asset problem specified as

$$\mu = \begin{bmatrix} 0.095 \\ 0.070 \\ 0.090 \\ 0.075 \end{bmatrix}, \qquad \Sigma = \begin{bmatrix} 0.0380 & 0.0085 & 0.0089 & 0.0066 \\ 0.0085 & 0.0331 & 0.0156 & 0.0039 \\ 0.0089 & 0.0156 & 0.0334 & 0.0070 \\ 0.0066 & 0.0039 & 0.0070 & 0.0240 \end{bmatrix}.$$

The efficient frontier for this example is plotted in Fig. 2.2, under the label "Efficient Frontier (no risk-free asset)".

2.3 Risk-Free Asset, Tangency Portfolio, Separation

When one of the assets can be considered risk-free (i.e. a return variance of zero and necessarily an identically zero covariance with all other assets), the above formulation cannot be used directly since the covariance matrix Σ would not be invertible. In this context, it can be shown that all efficient portfolios are formed by a linear combination of the risk-free asset and the *tangency portfolio* located on the risky-assets efficient frontier. These portfolios are located on what is known as the Capital Market Line (CML). These concepts, for a risk-free rate of 5%, are depicted on Fig. 2.2.

As derived in §A.2/p. 72, the risky-asset proportions of the tangency portfolio, given a risk-free rate R_f, are obtained as

$$\mathbf{w}^{\text{TGP}} = \frac{\Sigma^{-1}(\mu - R_f)}{\iota'\Sigma^{-1}(\mu - R_f)}.$$

A central consequence of the efficiency of all portfolios along the CML is that it is optimal for all investors (who share a common view about μ and Σ) to hold the *tangency portfolio in some proportion*. Investors only differ in their exposure to it, or alternatively, in how they allocate their holdings between the risk-free and tangency portfolio. This result was originally established by Tobin (1958) (see also Merton (1990, ch. 2)) and is an example of *separation* or *mutual fund* theorems.[1]

In the presence of a risk-free asset, portfolio optimization problems can be formulated without insisting on the "sum-to-one" constraint (2.4), since the unallocated fraction of capital, $1 - \mathbf{w}'\iota$, can be invested in the risk-free asset (or assumed to be borrowable at the risk-free rate in the case of a negative fraction).

Geometrically, from Fig. 2.2, the tangency portfolio can also be seen to maximize the *Sharpe ratio* (Sharpe 1966, 1994), defined as the expected portfolio excess return (over the risk-free rate R_f) per unit of portfolio return standard deviation,

$$\text{SR} \triangleq \frac{\mu_P - R_f}{\sigma_P},$$

with μ_P and σ_P given by eq. (2.1). A formal derivation of the relationship between the Sharpe ratio and the tangency portfolio appears in §A.2/p. 72.

2.4 Utility Maximization

Problem (2.2) does not specify what the "appropriate" level of target return ρ should be; this question should be decided by the investor and is a direct function of the risk

---

[1] This result also serves as a foundation for the celebrated Capital Asset Pricing Model (CAPM), which assumes, among other things, that all investors do share common views about μ and Σ, and examines equilibrium consequences; see §2.7.1/p. 23.

s/he is *willing* and *able* to bear. Markowitz (1959) introduces a formulation wherein
the investor's expected utility is directly maximized. He considered the following
quadratic form, written in terms of the portfolio return R_P,

$$U_\lambda(R_P) = R_P - \frac{\lambda}{2}R_P^2,$$

where λ is represents the investor's *risk aversion*, and in this context quantifies how
the investor is willing to trade each incremental unit of expected return against a
corresponding increase in variance of return.[2]

A rational decision maker would seek to maximize its *expected utility*, which is
computed as

$$\mathbb{E}[U_\lambda(R_P)] = \mu_P - \frac{\lambda}{2}\sigma_P^2$$
$$= \mathbf{w}'\mu - \frac{\lambda}{2}\mathbf{w}'\Sigma\mathbf{w},$$

where $\mathbf{w}$ is, as above, the weight given on each asset within the portfolio and μ_P and
σ_P^2 are respectively the mean and variance of the portfolio return distribution, given
by eq. (2.1). The expected quadratic utility maximization problem is then written as

$$\mathbf{w}^* = \arg\max_{\mathbf{w}} \mathbf{w}'\mu - \frac{\lambda}{2}\mathbf{w}'\Sigma\mathbf{w} \qquad (2.7)$$

$$\text{subject to} \quad \mathbf{w}'\iota = 1. \qquad (2.8)$$

When no further constraint is imposed, an analytical solution for $\mathbf{w}^*$ is easily found
by introducing Lagrange multipliers, similarly to the solution for problem (2.2).[3]

Proposition 1. *The unconstrained minimum-variance portfolio (2.2)–(2.3) and maximum quadratic utility (2.7) formulations are equivalent.*

Proof. The equality constraint (2.3) is incorporated in the minimum-variance objective (2.2) through an unconstrained Lagrange multiplier $v \in \mathbb{R}$, yielding the problem

$$\min_{\mathbf{w}} \frac{1}{2}\mathbf{w}'\Sigma\mathbf{w} - v(\mathbf{w}'\mu - \rho),$$

with first-order conditions for optimality given by $\Sigma\mathbf{w} - v\mu = 0$, yielding optimal
solution

$$\mathbf{w}^* = v\Sigma^{-1}\mu, \qquad (2.9)$$

where v is found by substitution as $v = \frac{\rho}{\mu'\Sigma^{-1}\mu}$.

---

[2] Many formulations of utility theory focus on the utility of *terminal wealth*, instead of the portfolio
return; Markowitz explicitly considers the latter (e.g. Markowitz 1959, p. 208), and this convention
is almost universally followed in mean-variance problems. An alternative formulation of quadratic
utility in terms of terminal wealth would slightly change the resulting equations.

[3] See, e.g. Chapados (2000) for a derivation.

Consider, on the other hand, the first-order optimality conditions of problem (2.7), $\mu - \lambda \Sigma \mathbf{w} = 0$, yielding optimal solution

$$\mathbf{w}^* = \frac{1}{\lambda} \Sigma^{-1} \mu. \tag{2.10}$$

Comparing eq. (2.9) and (2.10), it suffices to take $\lambda = \mu' \Sigma^{-1} \mu / \rho$ to obtain the equivalence. $\square$

This result confirms that in order to target a higher expected portfolio return ρ, the investor must exhibit a lower risk aversion.

Obviously, quadratic utility is but one of a number of utility functions that have been proposed to model the behavior of economic agents. The more general problem is easily written in terms of expected utility maximization,

$$\mathbf{w}^* = \arg\max_{\mathbf{w}} \int_{\mathbf{R}} U(\mathbf{w}'\mathbf{R}) \, dP(\mathbf{R}), \tag{2.11}$$

subject to the budget constraint (2.8), where $U(\cdot)$ is a utility function and $P(\mathbf{R})$ is the next-period return distribution. In particular, Mossin (1968) proves that constant relative risk aversion (CRRA) functions[4] are the only ones permitted if constant asset proportions are to be optimal, i.e. the investment in the risky asset does not depend on the level of initial wealth. Merton (1969) establishes the same result in a continuous-time setting. Moreover, Campbell and Viceira (2002) strongly argue in favor of CRRA utilities on the basis of the long-run observed behavior of the economy. However, for a large number of utility functions and "reasonable" return distributions, several studies (Levy and Markowitz, 1979; Kallberg and Ziemba, 1983) have established that single-period optimal portfolios under quadratic utility are very close to those obtained under alternative utilities.

A special case of some interest is the logarithmic utility, defined as $U(R) = \log(1 + R)$. This utility function is maximized by considering a Taylor series expansion of $1 + R$ around $R = 0$,

$$\log(1 + R) = R - \frac{R^2}{2} + O(R^3).$$

For relatively small returns, this is seen to be equivalent to the maximization of quadratic utility, problem (2.7), with $\lambda = 1$. The optimal weights under this utility function are given precisely by the tangency portfolio for a risk-free rate of zero (which also maximizes the Sharpe Ratio, see §A.2/p. 72). This property led some authors to confer a special aura to the logarithmic utility as being somehow "better",

---

[4] For a utility function $U(W)$, the Arrow–Pratt measure of relative risk aversion (Arrow, 1965; Pratt, 1964) is defined as

$$\text{RRA}(W) = -\frac{WU''(W)}{U'(W)}.$$

A CRRA utility function is one for which $\text{RRA}(W)$ is a constant independent of W. Such functions are sometimes said to exhibit *iso-elastic marginal utility*.

a point discussed, and found to be fallacious in a multiperiod setting, by Merton and Samuelson (1974). We return to the logarithmic utility in §3.4/p. 49.

Some utility functions have been proposed to incorporate parameter estimation uncertainty, the subject of *robust optimization*, which is covered in §2.8.5/p. 34.

2.5 Risk Measures

The exposition so far assumes that the investor considers the variance of the portfolio return distribution to be an adequate measure of risk. This measure has the major shortcoming that it considers positive return surprises to be as equally unpleasant as negative return surprises, a property that would surely be dismissed by most real-world investors! A number of alternative measures have been proposed throughout the years that attempt to quantify *portfolio downside risk*, starting with Markowitz's original treatment of the semivariance. This section briefly reviews the most significant possibilities. Nawrocki (1999) surveys the field more extensively.

2.5.1 Semivariance

Semivariance was originally considered by Markowitz (1959, Chapter 9) as a simple measure of downside risk. Whereas the variance is a symmetrical measure, semivariance only considers movements that fall below the mean; as such, its value depends on the *skewness* (third moment) of the distribution. For a scalar random variable X with mean μ, semivariance is defined as

$$\sigma^2_{\min} = \mathbb{E}\left[\min\left[X - \mu, 0\right]^2\right].$$

This measure can be used instead of portfolio variance in Problem (2.2). Although there is no closed-form solution to the mean-semivariance problem, Jin *et al.* (2006) establish the existence of the one-period mean-semivariance efficient frontier and review the literature examining its applications. Furthermore, Estrada (2007) provides an approximation to the semivariance that lends itself well to analytical solutions and reports good results on a number of problems.

2.5.2 Roy's Safety First

The Roy (1952) "safety-first" criterion puts portfolio risk in a more concrete setting than Markowitz' consideration of the second moment of returns. As Roy argued, the investor first decides on a minimum acceptable return that would ensure the preservation of a desired portion of his capital; he then proceeds with portfolio optimization by minimizing the probability of experiencing a return below the "disaster level". Let R_0 be the investor's minimum acceptable return and consider the problem

$$\text{minimize} \ \ P(R_P \leq R_0)$$
$$\text{subject to} \ \ \mathbf{w}'\iota = 1 \qquad \text{(budget)}.$$

Since the return distribution probability is not known precisely, this minimization may appear unfeasible. However, by Chebyshev's inequality, we have

$$P(R_P \leq R_0) \leq \frac{\sigma_P^2}{(\mu_P - R_0)^2},$$

which, taking square roots, yields the approximate problem

$$\min_{\mathbf{w}} \frac{\sigma_P}{\mu_P - R_0}$$

subject to the budget constraint. If the R_0 is the risk-free rate, this problem is equivalent to maximizing the Sharpe ratio (Sharpe, 1966).

2.5.3 Value-at-Risk

Value-at-Risk (VaR) was developed by JP Morgan in the early 1990's and made popular in a widely-circulated technical document (RiskMetrics, 1996) and associated software product. Intuitively, the level-α VaR (e.g. $\alpha = 95\%$) of a portfolio over a certain time horizon h is the portfolio return R_P such that the fraction α of returns will be better than R_P over the horizon. More formally, the level-α VaR of a portfolio is defined as the $1 - \alpha$-percentile of the portfolio return distribution,

$$\text{VaR}_\alpha(R_P) = -\inf_R \{R : P(R_P \geq R) \geq \alpha\},$$

where all returns are computed over horizon h. (The minus sign in the definition serves to make the risk measure positive.) The location of the VaR of an hypothetical asset return distribution, and its relationship to the CVaR (treated next) is shown in Fig. 2.3.

Value-at-Risk is regarded as a more plausible measure of portfolio risk than the variance since it accounts (in theory) for skewness and kurtosis in the return distribution.[5] In addition to its origins in risk management, it has received wide attention in a portfolio choice context where the VaR simply substitutes for the variance as the risk measure (Alexander and Baptista, 2002; Mittnik et al., 2003; Chow and Kritzman, 2002; Chapados, 2000).

---

[5] In practice, it is common to compute the VaR under a normal approximation due to its analytical tractability, which of course disregards higher-order moments in the underlying true distribution.

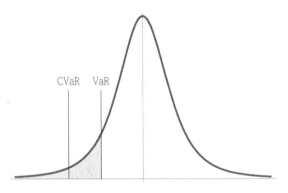

Fig. 2.3 90% Value-at-Risk (VaR) and Conditional Value-at-Risk (CVaR) for a Student $t(3)$ distribution. For fat-tailed distributions, the CVaR point can represent an expected loss much more significant than the VaR.

2.5.4 Conditional Value-at-Risk

In spite of its wide use, the VaR, as a measure of risk, suffers from a major defect: its lack of *subadditivity* (Artzner *et al.*, 1999). For a risk measure ρ applied to portfolios P_1 and P_2, subadditivity is satisfied if

$$\rho(P_1 + P_2) \leq \rho(P_1) + \rho(P_2),$$

which is a statement of the benefits of diversification—the risk of a diversified portfolio cannot be more than the risk of any of its constituents. That the VaR does not satisfy this property can lead to a number of counterintuitive results, particularly for firm-wide risk management, where it can appear that a more diversified portfolio exhibits a higher risk (Rau-Bredow, 2004).

A closely related measure that does satisfy subadditivity is the *conditional value at risk* (CVaR)—also called *expected shortfall* or *expected tail loss*—defined as the expected return conditional on observing a return lower than the VaR:

$$\mathrm{CVaR}_\alpha(R_P) = \mathbb{E}[R_P \,|\, R_P < \mathrm{VaR}_\alpha(R_P)],$$

where, as for the VaR, the returns are computed over a given time horizon h. In Fig. 2.3, this corresponds to an expectation taken within the shaded area. In a portfolio context, the CVaR has been studied by Krokhmal *et al.* (2002) and Consigli (2004).

2.5.5 Other Measures

In the past few years, there has been an explosion of alternative risk measures based on the modeling of tail phenomena (e.g. Malevergne and Sornette 2005a). Although it is not our focus to describe them in depth, Rachev *et al.* (2005) provide a good survey of the relevant literature, especially of measures related to portfolio selection.

Farinelli *et al.* (2006) provide computational portfolio allocation results comparing eleven alternative performance measure ratios.

2.6 Additional Constraints

Portfolio optimization problems, regardless of the form of the objective function or type of risk measure, are often solved with a number of constraints that attempt to capture *a priori* knowledge that the analyst possesses on what should be "good" solutions, embody investment objectives of the fund, or comply with regulatory requirements. It should be noted that with most of these constraints, Problem (2.2) can no longer be solved analytically but must instead be tackled with quadratic programming (Luenberger and Ye, 2007; Bertsekas, 2000) or mixed-integer programming (Wolsey and Nemhauser, 1999). Constraints also play a *regularization* role that can serve to mitigate sampling variance and estimation error in the mean return and risk forecasts; this is covered in §2.8/p. 29.

Some of the more common constraints are as follows. More comprehensive treatments appear in Fabozzi *et al.* (2006) and Qian *et al.* (2007). In line with the first reference, the rest of this section makes use of the following notation: we denote the current holdings of an investor by $\mathbf{w}_0$, the target holdings to be invested over the next period (i.e. the variables resulting from optimization) by $\mathbf{w}$, and their difference (the traded amount in each asset) by $\mathbf{x} = \mathbf{w} - \mathbf{w}_0$.[6] Furthermore, let $\mathbf{p}_0$ be the current price vector of the assets, and W_0 the current total portfolio value. The amount to be invested in asset i is given by $W_0 \mathbf{w}_i$ and the number of shares[7] held is $n_i = W_0 \mathbf{w}_i / \mathbf{p}_{0i}$.

2.6.1 No Short-Sales Constraint

This corresponds to the requirement that all portfolio weights be non-negative, namely

$$\mathbf{w}_i \geq 0, \quad \text{for all } i,$$

---

[6] The absolute traded amount, $|\mathbf{x}| = |\mathbf{w} - \mathbf{w}_0|$, shall be of significance, especially when considering transaction costs. The usual way of incorporating a term of this kind in a mathematical program is to introduce two variables,

$$\mathbf{x}^+ = \mathbf{w} - \mathbf{w}_0 \qquad \text{and} \qquad \mathbf{x}^- = \mathbf{w}_0 - \mathbf{w}$$

along with the constraints

$$\mathbf{x}^+ \geq 0 \qquad \text{and} \qquad \mathbf{x}^- \geq 0$$

and use the sum $\mathbf{x}^+ + \mathbf{x}^-$ whenever $|\mathbf{x}|$ appears.

[7] Assuming stocks as the assets.

thereby prohibiting selling assets short. Regulatory constraints placed on mutual
fund managers often mandate such a constraint. Markowitz' original formulation
of the portfolio choice problem included those constraints as an integral part of his
solution method, and many introductory treatments of the theory[8] include them by
default, despite the impossibility of deriving an analytical solution for the optimal
portfolio weights in their presence.[9]

2.6.2 Turnover and Transaction Costs Constraints

For large institutional portfolios, transaction costs can represent a sizable portion of
total operational costs, especially for funds that take an *active management* (Grinold
and Kahn, 2000) approach as opposed to a passive index-tracking objective. As
such, we may incorporate constraints that attempt to minimize the relative or dollar
turnover on individual assets, respectively

$$|\mathbf{x}_i| \leq U_i \qquad \text{and} \qquad W_0|\mathbf{x}_i| \leq \tilde{U}_i,$$

or the complete portfolio

$$\sum_i |\mathbf{x}_i| \leq U_P.$$

It is also possible to directly incorporate *transaction costs* into the objective function
as a term to be minimized. In its simplest form, a transaction cost model simply
imposes a proportional cost on the absolute value of traded quantities,

$$\text{prop cost}_i = W_0 \chi_i |\mathbf{x}_i|,$$

and the total portfolio cost given by

$$\text{prop cost}_P = W_0 \sum_i \chi_i |\mathbf{x}_i|, \tag{2.12}$$

where χ_i is the proportional cost of trading asset i. Assuming all χ_i and W_0 are non-
negative, prop cost$_P$ is nonnegative and hence the imposition of transaction costs
penalizes portfolio performance. To understand their consequence on realized returns, let $\mathbf{p}_1$ be the asset prices at the end of the investment period and consider the
relative return on asset i,

$$r_i = \frac{\mathbf{p}_{1i} - \mathbf{p}_{0i}}{\mathbf{p}_{0i}}.$$

Transaction costs affect portfolio return as

---

[8] E.g. Bodie *et al.* (2004).

[9] Non-negativity constraints can be seen as the "great divide" in optimization between analytical
and non-analytical solutions; in the case of portfolio optimization, the latter require, as mentioned
above, solution by quadratic programming.

$$\tilde{r}_i = \frac{\mathbf{p}_{1i} - \mathbf{p}_{0i} - W_0 \chi_i |\mathbf{x}_i|}{\mathbf{p}_{0i}}$$

$$= r_i - \frac{W_0}{\mathbf{p}_{0i}} \chi_i |\mathbf{x}_i| \quad = \quad r_i - n_i \chi_i |\mathbf{x}_i| \quad = \quad r_i - \tilde{\chi}_i |\mathbf{x}_i|,$$

where it is obvious that they adjust the portfolio relative return by a term proportional to the traded amount. Their effect can then directly be incorporated into the objective function for the quadratic utility maximization formulation, yielding the problem

$$\mathbf{w}^* = \arg\max_{\mathbf{w}} \mathbf{w}'\mu - \tilde{\chi}'|\mathbf{w} - \mathbf{w}_0| - \lambda \mathbf{w}' \Sigma \mathbf{w} \tag{2.13}$$

$$\text{subject to} \quad \mathbf{w}'\iota = 1. \tag{2.14}$$

The proportional costs structure is, however, only a starting point. As pointed out by Kissell and Glantz (2003), the totality of trading costs can be broken down according to an elaborate taxonomy that includes *explicit* (measurable) costs as well as more insidious *implicit* ones. Without delving into an intricate description, we can summarize them as follows:

Explicit Costs   They include *fixed costs*, in the form of commissions (as outlined above) and fees (custodial fees, transfer fees). They also include *variable costs*, in the form of bid–ask spread (the difference between the price at which one can buy versus sell) and taxes.[10]

Implicit Costs   They include *delay cost* (time between which a decision is made— for instance, by an allocation committee—and the actual trade is brought to the market), *price movement risk* (effect of underlying trends affecting the asset to be traded), *market impact costs* (deviation of the transaction price from the market price that would have prevailed had the trade not occurred), *timing risk* (cost attributable to general market volatility), *opportunity cost* (cost of not trading or not completing a trade).

Some of the implicit costs may not be costs at all but the source of trading profits depending on market conditions. A study by Wagner and Edwards (1998) shows that the price impact of a liquidity-demanding trade[11] averages −103 basis points[12] on a set of some 700,000 trades by more than 50 management firms in 1996, whereas the price impact of a liquidity-supplying trade generated *profits* of +36 basis points. In a liquidity-neutral market, the average price impact was −23 basis points. The effects of other implicit costs can likewise be decomposed according to market conditions.

There is a vast literature on transaction costs models, including how realistic nonlinear models of costs can be incorporated in asset-allocation models. This literature is well reviewed by Fabozzi *et al.* (2006, ch. 3).

---

[10] The proportional costs structure introduced previously can be seen as an adequate model of bid–ask spread, the most significant explicit cost for an institutional investor.

[11] For example, a "buy" trade executed when there are significantly more buyers than sellers.

[12] A *basis point* (bp) is one hundredth of one percent, i.e. $100\,\text{bp} = 1\%$.

2.6.3 Maximum Holdings Constraint

To ensure that the portfolio is not overly concentrated in a single asset, we can impose a constraint of the form

$$\mathbf{L} \leq \mathbf{w} \leq \mathbf{U},$$

where $\mathbf{L}$ and $\mathbf{U}$ are vectors specifying, respectively, the allowable lower and upper bounds for each asset. Likewise, we can ensure a sector $\mathscr{S} = \{i_1, i_2, \ldots, i_n\}$ (a set of asset indices) is not unduly weighted in the portfolio by imposing

$$L_{\mathscr{S}} \leq \sum_{i \in \mathscr{S}} \mathbf{w}_i \leq U_{\mathscr{S}},$$

with $L_{\mathscr{S}}$ and $U_{\mathscr{S}}$ denoting, respectively, the minimum and maximum exposure to the sector.

2.6.4 Maximum Tracking Error and Factor Exposure Constraint

The performance of portfolio managers is often compared to that of a *benchmark* such as the S&P 500 (Grinold and Kahn, 2000). Depending on the fund's style, the manager may seek to replicate the benchmark as closely as possible (using, for instance, a smaller number of assets than the benchmark), or to provide additional performance (the so-called "alpha") at the expense of taking on *active risk*, namely, deviating from the benchmark. This risk is quantified by the *tracking error*, defined next. Assume that the benchmark's and fund's investable universe are the same and that the (random) asset returns are given by $\mathbf{R}$. Let $\mathbf{w}_b$ denote the benchmark weights, $\mathbf{w}$ the decision variables, and R_B and R_P denote, respectively, the benchmark and portfolio returns,

$$R_B = \mathbf{w}_B'\mathbf{R} \qquad \text{and} \qquad R_P = \mathbf{w}'\mathbf{R}.$$

The tracking error is simply the variance of the return difference between the benchmark and the invested portfolio,[13]

$$
\begin{aligned}
\text{TE}_P &= \text{Var}[R_P - R_B] \\
&= \text{Var}[\mathbf{w}_B'\mathbf{R} - \mathbf{w}'\mathbf{R}] \\
&= (\mathbf{w}_B - \mathbf{w})'\Sigma(\mathbf{w}_B - \mathbf{w}),
\end{aligned}
$$

with Σ the asset return covariance matrix. A quadratic tracking error constraint of the form

---

[13] More accurately, *tracking error* is usually reserved for the square-root of this variance, but for notational simplicity, we shall omit the square-roots in this overview.

$$(\mathbf{w}_B - \mathbf{w})'\Sigma(\mathbf{w}_B - \mathbf{w}) \leq \sigma_{\text{TE}}^2$$

can then be imposed to limit active risk. Note that this does not limit *total risk*, which would require additional constraints (Jorion, 2003).

In an analogous manner, one can restrict exposure to specific risk factors. Suppose that we posit the following decomposition for *explaining* the return of asset i as a linear combination of factors (additional background on factor models is given in §2.7/p. 23),

$$R_i = \alpha_i + \sum_{j=1}^{M} \beta_{i,j} F_j + \varepsilon_i,$$

where F_j is the random "return" associated with factor j[14] during the period, and $\beta_{i,j}$ is the exposure of asset i to factor j. This is written more succinctly as

$$\mathbf{R} = \alpha + \mathbf{BF} + \varepsilon$$

with $\mathbf{B}$ and $\mathbf{F}$ respectively the matrix of factor exposures and the vector of one-period factor returns. This yields a portfolio return, given asset weights $\mathbf{w}$, of

$$R_P = \mathbf{w}'\alpha + \mathbf{w}'\mathbf{BF} + \mathbf{w}'\varepsilon.$$

The exposure of the portfolio to factor j is given by $\sum_i \mathbf{w}_i \beta_{i,j}$. Bound or equality constraints may be placed on this exposure; for example, to ensure an *ex ante* neutral exposure to factor j one may impose

$$\sum_i \mathbf{w}_i \beta_{i,j} = 0.$$

Such constraints are commonly used in so-called "long-short equity" hedge funds, which are designed to be neutral to overall market fluctuations.[15]

2.6.5 Transaction Size, Cardinality and Round Lot Constraints

The following class of constraints is of a combinatorial nature and necessitates solution by *mixed integer programming* methods (Wolsey and Nemhauser, 1999). For convenience, we define the vector δ of binary indicator variables

$$\delta_i = \begin{cases} 1, & \text{if } \mathbf{w}_i \neq 0, \\ 0, & \text{if } \mathbf{w}_i = 0, \end{cases} \quad i = 1,\ldots,N,$$

[14] For stocks, examples of likely factors would be the return on a broad market index, the return difference between growth and value stocks, and the return difference between large- and small-capitalization stocks; see §2.7/p. 23.

[15] For a factor-neutral constraint to make sense, the exposures $\beta_{i,j}$ must be standardized to have a mean of zero across assets.

where each element specifies whether a position is being taken in the corresponding asset.

A first class of combinatorial constraints aims at eliminating positions that are too small; such positions are often the result of a traditional unconstrained mean–variance optimization. The manager can require

$$|\mathbf{w}_i| \geq \delta_i \mathbf{L}_{\mathbf{w}_i},$$

where $\mathbf{L}_{\mathbf{w}_i}$ is the minimum (relative) position size allowed for asset i. Likewise a limit can be set on portfolio trades

$$|\mathbf{x}_i| \geq \delta_i \mathbf{L}_{\mathbf{x}_i},$$

with $\mathbf{L}_{\mathbf{x}_i}$ the minimum allowed trade size for asset i.

Next, *cardinality* constraints can be useful in problems that seek to replicate a benchmark using a smaller number of assets than the original universe. This may take the form of

$$\delta' \iota \leq K$$

where K is the maximum number of allowable assets. The impact of cardinality constraints on the shape of the efficient frontier is studied by Chang *et al.* (2000).

Finally, *round lot* constraints account for the fact that market-traded instruments are not infinitely divisible (contrarily to idealizations of finance theory)—it is common for stocks to be traded in multiples of 100 shares or more. If the lot size for asset i is given by the constant κ_i and the desired number of lots by η_i (an integer decision variable), we can enforce

$$W_0 \mathbf{w}_i = \kappa_i \eta_i \mathbf{p}_{0i}, \qquad \eta_i \in \mathbb{Z}.$$

In general, when imposing round lots, the budget constraint, $\sum_i \mathbf{w}_i = 1$, may no longer be satisfiable; in this case, one may settle for an approximate budget constraint, expressed as

$$\frac{1}{W_0} \sum_i \kappa_i \eta_i \mathbf{p}_{0i} + \xi^+ - \xi^- = 1,$$
$$\xi^+, \xi^- \geq 0,$$
$$\eta_i \in \mathbb{Z},$$

where ξ^+ and ξ^- are "slack variables" to be minimized (by incorporating them in the objective function). Formulations of this type are analyzed by Kellerer *et al.* (2000).

2.7 Forecasting Models

Markowitz's method of portfolio construction is silent on how the required expected next-period asset returns and covariances are to be obtained. This section reviews the most commonly-used approaches in practice, starting with *factor models* and their uses in covariance modeling and expected return forecasts. We then briefly cover other expected-return forecasting approaches for equities, mostly based on *dividend discount models* and accounting ratios. Finally, extensive experience with mean-variance criteria suggest that they are extremely sensitive to parameter estimation error—very small changes in the forecasts can yield enormous changes in "optimal" portfolio weights, leading to doubt about the validity of the portfolios and possible considerable rebalancing costs when the decisions are implemented. This naturally paves the way for robust estimation methods and Bayesian approaches; we cover some of the methods that have been suggested to counter portfolio instability.

2.7.1 Factor Models

Factor models seek to explain the *cross-section*[16] of asset returns by a simple affine relationship, where the return of asset i over the period t is decomposed into the return of more elemental *factor returns* $F_{j,t}$,

$$R_{i,t} = \alpha_i + \sum_{j=1}^{M} \beta_{i,j} F_{j,t} + \varepsilon_{i,t}, \qquad (2.15)$$

where α_i is a regression constant, $\beta_{i,j}$ are the factor exposures, and $\varepsilon_{i,t}$ is a zero-mean random unexplained component uncorrelated with factor returns.[17]

The grandfather of factor models is the Capital Asset Pricing Model (CAPM) of Sharpe (1964), Lintner (1965) and Mossin (1966); this model is generally derived from equilibrium considerations as a *positive theory* of collective investor behavior,[18] but we shall merely regard it as a simple one-factor model. It expresses the expected excess return[19] on asset i as a linear function of the return of the overall market portfolio, R_M,

$$\mathbb{E}[R_i - R_f] = \beta_i \mathbb{E}[R_M - R_f],$$

---

[16] As opposed to the time-series characteristics.

[17] It should be noted that what this literature refers to as *factors* almost exclusively consist of observable variables, what would simply be called covariates, explanatory or input variables in a more traditional statistical context. Latent factors are always referred to as such.

[18] In other words, it seeks to establish what consequences would arise if every investor behaved according to a set of hypotheses that include Markowitz's rules for portfolio choice among others.

[19] The return earned over the risk-free rate.

where, under the CAPM assumptions, α_i is identically zero.[20]

It has long been understood, at least since Merton (1973), that there exists the possibility that additional sources of *priced risk*, on top of the market portfolio, could impact expected asset returns. Generalizations of the CAPM are obtained in the context of the Arbitrage Pricing Theory (APT) of Ross (1976).[21] Assume that asset returns are distributed according to the factor structure of eq. (2.15), along with

$$\mathbb{E}[\varepsilon_i] = \mathbb{E}[F_k] = 0$$
$$\mathbb{E}[\varepsilon_i\varepsilon_j] = \mathbb{E}[\varepsilon_iF_j] = \mathbb{E}[F_iF_j] = 0, \quad i \neq j$$
$$\mathbb{E}[\varepsilon_i^2] = \sigma^2 < \infty.$$

In this context, in the absence of arbitrage and under some technical conditions, Ross showed that the excess return on asset i is given by

$$\mathbb{E}[R_i - R_f] = \sum_{j=1}^{K} \beta_{i,j}\mathbb{E}[F_j - R_f].$$

Under the APT, each factor represents a priced systematic risk (a risk for which investors are seeking compensation), and the factor exposures $\beta_{i,j}$ quantify the *market price* of those risks (how much the investor is compensated in expected return for taking on a unit of risk).

Ross remains silent on how factors should be chosen. In addition to the CAPM market portfolio factor, several *pricing anomalies* have been documented in the 1980's and early 1990's suggesting additional factors, including long-run price reversal (De Bondt and Thaler, 1985), short-run price momentum (Jegadeesh and Titman, 1993), and a variety of effects due to firm size (market equity, ME, the stock price times the number of shares), earnings to price ratio (E/P), cash-flow to price ratio (C/P), book value to market value (BE/ME), and past sales growth (Banz, 1981; Basu, 1983; Rosenberg *et al.*, 1985; Lakonishok *et al.*, 1994). These results built up to an influential series of papers by Fama and French (1992; 1993; 1995; 1996), who show that the following two additional factors summarize well a number of empirical findings:

High-Minus-Low   (HML) The difference between the return on a portfolio of high-book-to-market stocks and the return on a portfolio of low-book-to-market stocks.[22]

---

[20] Starting from the late-1960's, a huge literature has emerged aiming at testing the validity of the CAPM; see Campbell *et al.* (1997) for an overview.

[21] Technically, the CAPM is derived from equilibrium considerations whereas the APT is derived from a more fundamental "absence of arbitrage" principle; these minutiæ make little difference from a statistical estimation standpoint.

[22] The precise definition is slightly technical and appears in Fama and French (1996).

Small-Minus-Big   (SMB) The difference between the return on a portfolio of small-capitalization stocks and the return on a portfolio of large-capitalization stocks.

Put together, Fama and French argue that a model of the form

$$\mathbb{E}[R_i - R_f] = \beta_i \mathbb{E}[R_M - R_f] + s_i \mathbb{E}[\text{SMB}] + h_i \mathbb{E}[\text{HML}]$$

can account for a large fraction of the cross-section of returns, and obtain times-series regression R^2 in the 0.90–0.95 range. The only factor significantly unaccounted for is the short-run price momentum, which is empirically analyzed by Carhart (1997).

Since the late 1990's, several large commercial factor models have become available, the best known of which is perhaps Barra's fundamental multifactor risk model for United States equities (Barra, 1998), which includes 13 risk indices and 55 industry groups.

2.7.2 Factor Models in Covariance Matrix Estimation

The estimation of covariance matrices for portfolios of many assets is a hard problem. As an illustration, consider the Russell 1000 index, whose sample covariance matrix

$$\hat{\Sigma} = \frac{1}{T-1} \sum_{t=1}^{T} (\mathbf{R}_t - \hat{\mu})(\mathbf{R}_t - \hat{\mu})'$$

contains 500,500 distinct entries;[23] an analysis with the tools of *random matrix theory* shows that for such large matrices, only a few eigenvalues of the sample covariance matrix carry information, the rest being the result of noise (Laloux *et al.*, 1999; Malevergne and Sornette, 2005b). This observation gave rise to a number of schemes to add structure to the estimator, often relying on *shrinkage methods* that attempt to find an optimal compromise between a restricted and unrestricted estimators (§2.8/p. 29).

An obvious application of factor models is to the estimation of covariance matrices. This approach can be traced back to a suggestion by Sharpe (1963), and relies on the factor decomposition of eq. (2.15). Assume that firm-specific residual returns, ε_i, are uncorrelated for two different firms,

$$\mathbb{E}[\varepsilon_i \varepsilon_j] = \begin{cases} 0, & i \neq j, \\ \sigma_i^2, & i = j. \end{cases}$$

The covariance between returns R_i and R_j is obtained from eq. (2.15) as

---

[23] Obtained as $1000 \times 1001/2$.

$$\mathrm{Cov}[R_i, R_j] = \sum_{k=1}^{M} \mathrm{Cov}[\beta_{i,k} F_i, \beta_{j,k} F_j] + \mathrm{Cov}[\varepsilon_i, \varepsilon_j]$$

$$= \sum_{k=1}^{M} \beta_{i,k} \beta_{j,k} \mathrm{Cov}[F_i, F_j] + \delta_{i,j} \sigma_i^2,$$

where $\delta_{i,j}$ is the Kronecker delta. This expression illustrates that under a factor model of returns, the covariance between arbitrary assets depends only on the *co-variance matrix between the individual factors*, which (for the small number of factors used in practice) is a much more tractable quantity to estimate with statistical reliability. Current methods for covariance modeling are reviewed by Fabozzi *et al.* (2006) and Qian *et al.* (2007).

2.7.3 Factor Models in Expected Return Estimation

Forecasting expected asset returns is recognized as notoriously difficult — so much so that this apparent unforecastability gave rise to the Efficient Market Hypothesis (EMH) and a famous proof that prices should fluctuate randomly (Cootner, 1964; Samuelson, 1965; Fama, 1970). Empirically, it is often observed that the simplest predictors, a constant based on the historical average return or even the constant *zero*,[24] perform the best out of sample. More recently, with advances in computing power and improvements in the quality and quantity of available data, mounting evidence has started to accumulate in favor of some *very small* forecastability (Lo and MacKinlay, 1999), possibly arising from market imperfections. However, exploiting any residual forecastability, especially when accounting for trading costs, remains of the utmost challenge.

Factor models can provide some direction in this respect and are generally used by relating the returns at time t with the observed factors at the same time, and then positing a dynamical model for making forecasts of the factors themselves. It is common to utilize a Vector Autoregressive (VAR) model for establishing the dynamics (Hamilton, 1994), yielding an overall forecasting model specified as

$$\mathbf{R}_t = \alpha + \beta' \mathbf{F}_t + \varepsilon_{\mathbf{R},t}$$
$$\mathbf{F}_{t+1} = \mathbf{a} + \mathbf{B}\mathbf{F}_t + \varepsilon_{\mathbf{F},t+1},$$

where $\mathbf{a}$ is a vector and $\mathbf{B}$ is a matrix of first-order autoregression factors.

An example that has received wide attention is the forecastability of stock returns by the dividend yield.[25] Brandt (2004) estimates the following parameters for the

---

[24] Which is surprisingly effective in the case of daily stock returns.

[25] The first evidence is presented in Campbell and Shiller (1988) and Fama and French (1988); Campbell (1991) presents an interesting decomposition of stock returns wherein he shows that unexpected stock returns must be associated with changes in expected future dividends or expected future returns, and attributes a third of the variance in U.S. unexpected returns over the 1927–88

quarterly returns of the value-weighted CRSP[26] index

$$
\begin{bmatrix} r^e_{t+1} \\ d_{t+1} - p_{t+1} \end{bmatrix} = \begin{bmatrix} \underset{(0.0839)}{0.2049} \\ \underset{(0.0845)}{-0.1694} \end{bmatrix} + \begin{bmatrix} \underset{(0.0249)}{0.0568} \\ \underset{(0.0251)}{0.9514} \end{bmatrix} (d_t - p_t) + \begin{bmatrix} \varepsilon_{1,t+1} \\ \varepsilon_{2,t+1} \end{bmatrix} \tag{2.16}
$$

$$
\begin{bmatrix} \varepsilon_{1,t+1} \\ \varepsilon_{2,t+1} \end{bmatrix} \sim N\left(\begin{bmatrix} 0 \\ 0 \end{bmatrix}, \begin{bmatrix} 0.0062 & -0.0060 \\ -0.0060 & 0.0063 \end{bmatrix} \right),
$$

where r^e_t denotes the log excess return of the index and $d_t - p_t$ is the log dividend yield, computed from the log of the trailing-twelve-month sum of monthly dividends d_t and the current index level p_t.[27] In parenthesis are the Newey and West (1987) standard errors. These results serve to illustrate that whatever forecastability remains, although statistically significant over a long sample, remains low.

2.7.4 Other Expected Return Forecasting Models

A different angle on forecasting models for equities is provided by the *fundamental analysis* of a firm's fair value. The starting point in this line of study is the *dividend discount model* (DDM), introduced by Williams (1938), stating that the price of one share of stock should be given by the sum of discounted future dividend payments,

$$
P_t = \mathbb{E}_t \left[\sum_{\tau=1}^{\infty} \frac{D_{t+\tau}}{(1+R_{t+\tau})^\tau} \right], \tag{2.17}
$$

where D_t is the dividend to be paid in (future) period t and R_t are discount rates.[28] It should be noted that the discount rate is generally higher than the prevailing risk-free rate and reflects the market's expectations on the prospects of future dividend payments; a greater risk on the dividend stream entails a higher discount rate. In other words, it can be viewed as the rate of return that investors *require* for bearing the risk of holding the equity. Consider a simplification wherein we keep the discount factor constant (i.e. not time-varying, but still unknown) with value R and assume a constant growth rate g for dividends,[29] $D_{t+1} = D_t(1+g) = D_1(1+g)^{t-1}$,

---

period to the first component, a third to the second, and the final third to their covariance. For use of the dividend yield in an asset allocation context, see e.g. Kandel and Stambaugh (1996) and Brennan *et al.* (1997).

[26] Center for Research in Security Prices, based at the University of Chicago; www.crsp.com.

[27] The estimation period in this example is from April 1952 to December 1996, and the results are fairly stable across different estimation periods.

[28] This model can be adapted to a similar *free cash flow* relationship for stocks that do not pay dividends.

[29] This hypothesis is valid, for instance, under the scenario where a business grows its earnings at a constant rate and maintains the same dividend payout ratio.

which allows to write

$$P_t = \mathbb{E}_t \left[\sum_{\tau=1}^{\infty} \frac{D_{t+\tau+1}(1+g)^{\tau-1}}{(1+R)^{\tau}} \right] = \mathbb{E}_t \left[\frac{D_{t+1}}{R-g} \right].$$

This is referred to as the Gordon (1962) growth model. Now assuming that price P_t is observed on the market and that R independent of D_{t+1} (the latter is generally a quite well ascertained quantity), the expected implied discount rate—thence the implied expected return on the security—can be solved for as

$$\mathbb{E}_t[R] = \frac{\mathbb{E}_t[D_{t+1}]}{P_t} + g.$$

Unfortunately, this model is very sensitive to inaccuracies in its inputs, and for this reason, so-called *residual income valuation* models (RIM) have been proposed that exploit the fundamental accounting *clean surplus relationship* linking the balance sheet and income statement

$$B_t = B_{t-1} + E_t - D_t, \tag{2.18}$$

where B_t is the firm's book value per share at time t and E_t the earnings per share generated during period t. This states that the period-to-period variation in the firm's value is given by increases resulting from the period activities (net earnings) minus payments to shareholders (dividends) (Edwards and Bell, 1961; Ohlson, 1995). Define the "abnormal" earnings, assuming a constant discount factor R, as

$$E_t^a \triangleq E_t - R B_{t-1};$$

in this context, R can be interpreted as the required return on equity expected at the start of each period. This relationship, in conjunction with eq. (2.18), allows to write the dividends for period t as

$$D_t = E_t^a - B_t + (1+R) B_{t-1}.$$

Substituting in eq. (2.17), we obtain

$$
\begin{aligned}
P_t &= \mathbb{E}_t \left[\frac{D_{t+1}}{1+R} + \frac{D_{t+2}}{(1+R)^2} + \cdots \right] \\
&= \mathbb{E}_t \left[\frac{E_{t+1}^a - B_{t+1} + (1+R) B_t}{1+R} + \frac{E_{t+2}^a - B_{t+2} + (1+R) B_{t+1}}{(1+R)^2} + \cdots \right] \\
&= B_t + \mathbb{E}_t \left[\sum_{\tau=1}^{\infty} \frac{E_{t+\tau}^a}{(1+R)^{\tau}} \right] \\
&= B_t + \mathbb{E}_t \left[\sum_{\tau=1}^{\infty} \frac{E_{t+\tau} - R B_{t+\tau-1}}{(1+R)^{\tau}} \right].
\end{aligned}
$$

Under some assumptions, Philips (2003) derives the following expression for the expected returns

$$\mathbb{E}_t[R] = \frac{\mathbb{E}_t[E_{t+1}] - gB_t}{P_t} + g,$$

where P_t and B_t are readily available and E_{t+1} is often estimated by analysts that follow a stock.[30] The growth rate g can conservatively be taken as the growth of nominal GDP.[31] Claus and Thomas (2001) find relationships based on residual income valuations to be much less sensitive to errors than the Gordon model.

The topic of expected return forecasts is much richer than this brief overview can provide. In particular, we must omit treatment of a sizable literature on the information regarding the implied probability distribution of returns that option markets provide (e.g. Pan and Poteshman 2006; Aït-Sahalia and Brandt 2007). A review of several recently-proposed methodologies for forecasting expected returns appears in Satchell (2007).

2.8 Forecast Stability and Econometric Issues

A longstanding critique of Markowitz's mean-variance method of portfolio choice stems from the often-observed erratic nature of the optimal weights: unless expected returns are "perfectly matched" to the covariance matrix, it is frequent to arrive at *corner solutions* wherein a small number of assets get allocated most of the weight, with problem constraints strongly governing the obtained solution. It almost appears as if the theory's foundational goal of *efficient diversification of investment*[32] somehow gets lost along the way. Moreover, the obtained solutions tend to be unstable, both cross-sectionally (small changes to the forecasts have a large impact on the weights) and over time (optimal portfolios often change drastically from one period to the next, leading to important costs due to turnover).

Michaud (1989) argues that extreme and unstable portfolio weights are inherent to mean-variance optimizers due to forecast estimation error: by virtue of mere statistical fluctuation, large positive (negative) weights are assigned to assets that have large positive (negative) estimation error in expected return and/or large negative (positive) error in variance. This arises because in the classical mean-variance

---

[30] Analyst forecasts of earnings have themselves long been subject to investigation, including the early work of Crichfield *et al.* (1978) and Givoly and Lakonishok (1984), who generally find forecasts to improve as the earnings publication date approaches. More recently, Friesen and Weller (2006) consider a Bayesian framework in which analysts constantly revise their forecasts based on newly-revised information; in this context, the authors report strong evidence of biases, including overconfidence and cognitive dissonance biases.

[31] For firms whose capital structure consists of a mixture of equity and debt, this is indeed a very conservative assumption. The growth rate of nominal GDP would normally characterize the return on the firm's *assets*. In contrast, the return on *equity*—the quantity represented by g—would be magnified by the firm's financial leverage, i.e. its use of debt.

[32] The subtitle in Markowitz's 1959 treatment of the subject.

paradigm, forecasts are totally disconnected from optimization: the former are "plugged into" the latter (hence the name *plug-in estimates*), and in a sense the optimizer "does not know" that the forecasts are but point estimates that also have an associated standard error. This led Michaud to his bon mot that *mean-variance optimizers act as statistical error maximizers.*

Michaud (1989), Jobson and Korkie (1980), Best and Grauer (1991) and Chopra and Ziemba (1993) study the impact of estimation uncertainty, where it is often observed to be much larger than that of asset risk itself. In particular, the plug-in estimates are found to be extremely unreliable, their performance dropping rapidly as the number of assets increases. This led to a variety of approaches to "robustify" the optimal portfolios, including shrinkage estimators, Bayesian approaches, resampling methods and robust optimization, summarized next. It should be noted that the practitioner's little-told secret of imposing optimization constraints, such as those reviewed in §2.6/p. 17, already serves to stabilize the portfolio by truncating extreme weights, and was confirmed by Frost and Savarino (1988) to generally improve performance. In this context, constraints can be interpreted as providing a *post hoc* regularization of the estimator, a point elaborated upon by Jagannathan and Ma (2003).

A very complete review of the literature on the econometrics of portfolio choice appears in Brandt (2004).

2.8.1 Shrinkage Estimators

It is known since Stein (1956) that biased estimators often have better finite-sample properties (lower sample variance) than unbiased ones.[33] In particular, consider estimating the mean of an N-dimensional ($N \geq 3$) multivariate normal distribution with *known covariance matrix* Σ, subject to the quadratic loss function

$$L(\hat{\mu}, \mu) = (\hat{\mu} - \mu)' \Sigma^{-1} (\hat{\mu} - \mu),$$

where μ is the true mean. In this context, the usual sample mean $\hat{\mu}$ is not the best estimator (James and Stein, 1961). The James-Stein *shrinkage estimator*

$$\hat{\mu}_{JS} = (1-w)\hat{\mu} + w\mu_0 \iota, \qquad 0 < w < 1,$$

exhibits a lower quadratic loss, where μ_0 is an arbitrary "common" constant and is called the shrinkage target. The optimal trade-off between bias and variance is achieved by

$$w^* = \min\left(1, \frac{(N-2)/T}{(\hat{\mu} - \mu_0 \iota)' \Sigma^{-1} (\hat{\mu} - \mu_0 \iota)}\right).$$

---

[33] This *bias–variance trade-off* is related to the notion of capacity control which is studied in depth in machine learning; see, e.g. Bishop (2006) and Hastie *et al.* (2009) for textbook treatments.

More generally, shrinkage methods involve the combination of an unstructured estimator (with a large number of degrees of freedom and likely high sample variance)
and a highly structured one (with a small number or even zero degrees of freedom).
Jobson and Ratti (1979) and Jorion (1986) have studied them in a portfolio context,
demonstrating that their benefits carries to the estimation of expected returns and
obtain good performance of the resulting portfolios. Similarly, Frost and Savarino
(1986) and Ledoit and Wolf (2004) apply them to the estimation of covariance matrices. Brandt (2004) suggests applying shrinkage estimation directly to the optimal portfolio weights, where the shrinkage target can be some *ex ante* reasonable
weights such as $1/N$ or those of a benchmark portfolio.

2.8.2 Bayesian Approaches

In contrast to the "plug-in" approaches presented previously which sought to obtain
the single best estimates of the next-period return mean and variance, a Bayesian or
decision-theoretic approach would explicitly carry the estimation uncertainty to the
optimization. Consider an explicit parametrization of the next-period return distribution, $P(\mathbf{R} \mid \theta)$, in terms of a parameter vector θ, allowing us to rewrite the expected
utility maximization, eq. (2.11), as

$$\mathbf{w}^*(\theta) = \arg\max_{\mathbf{w}} \int_{\mathbf{R}} U(\mathbf{w}'\mathbf{R}) \, dP(\mathbf{R} \mid \theta).$$

A Bayesian investor would not commit to a single choice of parameter vector θ,
but would instead consider the posterior distribution of parameters, given by Bayes'
rule as

$$P(\theta \mid \mathcal{D}) = \frac{P(\mathcal{D} \mid \theta) P_0(\theta)}{P(\mathcal{D})},$$

where $\mathcal{D}$ is some data (obviously only known up to before the start of the forecast period) and $P_0(\theta)$ is a (subjective) prior distribution on parameter values. The
investor's subjective distribution of asset returns, given the data, is obtained by
marginalizing out the parameters,

$$P(\mathbf{R} \mid \mathcal{D}) = \int_{\theta} P(\mathbf{R} \mid \theta) \, dP(\theta \mid \mathcal{D}),$$

yielding to reformulating the expected utility maximization problem for finding optimal portfolio weights as

$$\mathbf{w}^* = \arg\max_{\mathbf{w}} \int_{\theta} \left[\int_{\mathbf{R}} U(\mathbf{w}'\mathbf{R}) \, dP(\mathbf{R} \mid \theta) \right] dP(\theta \mid \mathcal{D}).$$

This approach to portfolio choice was pioneered as early as the 1960's by Zellner
and Chetty (1965) and further studied by Klein and Bawa (1976) and Brown (1978).

More recently, the notion of a "learning investor" was revisited in the context of the increasing evidence on the (mild) predictability of returns in works by Kandel and Stambaugh (1996) and Barberis (2000); see §3.5/p. 55.

2.8.3 The Black-Litterman Model

A different path to Bayesian estimation relies on the implications of an underlying economic equilibrium model, which can serve to provide the "prior" in a portfolio choice context. This is embodied in the Black and Litterman (1992) model, widely used by practitioners. Our presentation of this model draws from Fabozzi *et al.* (2006).

Consider the expected-return relationship for asset i given by the CAPM (§2.7.1/p. 23),

$$\Pi_i = \mathbb{E}[R_i - R_f] = \beta_i \mathbb{E}[R_M - R_f], \tag{2.19}$$

where β_i is obtained as a regression coefficient,

$$\beta_i = \frac{\text{Cov}[R_i, R_M]}{\sigma_M^2},$$

with σ_M^2 the variance of the market portfolio. We shall denote by $\mathbf{w}_M$ the weights of the market portfolio, such that its return can be written as

$$R_M = \sum_{j=1}^{N} \mathbf{w}_{M,j} R_j.$$

Then eq. (2.19) can be rewritten as

$$
\begin{aligned}
\Pi_i &= \beta_i \mathbb{E}[R_M - R_f] \\
&= \frac{\text{Cov}[R_i, R_M]}{\sigma_M^2} \mathbb{E}[R_M - R_f] \\
&= \frac{\text{Cov}[R_i, \sum_{j=1}^{N} \mathbf{w}_{M,j} R_j]}{\sigma_M^2} \mathbb{E}[R_M - R_f] \\
&= \frac{\mathbb{E}[R_M - R_f]}{\sigma_M^2} \sum_{j=1}^{N} \mathbf{w}_{M,j} \text{Cov}[R_i, R_j],
\end{aligned}
$$

or in matrix form,

$$\Pi = \delta \Sigma \mathbf{w}_M \qquad \text{with} \quad \delta = \frac{\mathbb{E}[R_M - R_f]}{\sigma_M^2}.$$

Although the true expected asset returns μ are unknown, we can posit that the equilibrium model provides a sensible approximation in the form of

$$\Pi = \mu + \varepsilon_\Pi, \qquad \varepsilon_\Pi \sim N(0, \tau\Sigma), \tag{2.20}$$

where $\tau \ll 1$ is a small constant.[34] We can view ε_Π as a "confidence interval" in which the true expected returns are approximated by the equilibrium model: a small τ implies a high confidence in the equilibrium estimates and vice versa.

Now suppose that the investor holds particular *views* on some assets or combinations of assets; examples are "the expected return of asset i will be x percent", or "asset j will outperform asset k by z percent". Each view has an attached *confidence* reflecting how strongly the investor believes them. We can formally express the K views as a vector $\mathbf{q} \in \mathbb{R}^K$,

$$\mathbf{q} = \mathbf{P}\mu + \varepsilon_\mathbf{q}, \qquad \varepsilon_\mathbf{q} \sim N(0, \Omega), \tag{2.21}$$

where $\mathbf{P}$ is a $K \times N$ matrix of view combinations and Ω is a $K \times K$ matrix of view confidences. For example, in a universe of $N = 3$ assets, the investor may believe that

- Asset 1 will have a return of 1.5%.
- Asset 3 will outperform asset 2 by 4%.

This yields the following form for the views

$$\begin{bmatrix} 1.5\% \\ 4\% \end{bmatrix} = \begin{bmatrix} 1 & 0 & 0 \\ 0 & -1 & 1 \end{bmatrix} \begin{bmatrix} \mu_1 \\ \mu_2 \\ \mu_3 \end{bmatrix} + \begin{bmatrix} \varepsilon_{\mathbf{q},1} \\ \varepsilon_{\mathbf{q},2} \end{bmatrix},$$

for some view confidence matrix Ω, which is commonly diagonal. Both eq. (2.20) and (2.21) are expressed in terms of the unknown expected returns μ. The Black-Litterman model uses the *mixed estimator* of Theil and Goldberger (1961) to combine the information from two data sources—here the equilibrium model and the investor views—into a single posterior estimator. Start by "stacking" the two equations as follows,

$$\mathbf{y} = \mathbf{X}\mu + \varepsilon, \qquad \varepsilon \sim N(0, \mathbf{V})$$

where

$$\mathbf{y} = \begin{bmatrix} \Pi \\ \mathbf{q} \end{bmatrix}, \qquad \mathbf{X} = \begin{bmatrix} \mathbf{I}_N \\ \mathbf{P} \end{bmatrix}, \qquad \mathbf{V} = \begin{bmatrix} \tau\Sigma & \\ & \Omega \end{bmatrix}.$$

We can rely on a standard generalized least squares (GLS) estimator (Greene, 2007) to arrive at the *Black-Litterman* estimator for expected returns,

---

[34] Values in the neighborhood of 0.1–0.3 often give satisfactory results for U.S. equities.

$$
\begin{aligned}
\hat{\mu}_{\text{BL}} &= (\mathbf{X}'\mathbf{V}^{-1}\mathbf{X})^{-1}\mathbf{X}'\mathbf{V}^{-1}\mathbf{y} \\
&= \left(\begin{bmatrix} \mathbf{I}_N \, \mathbf{P}' \end{bmatrix} \begin{bmatrix} (\tau\Sigma)^{-1} & \\ & \Omega^{-1} \end{bmatrix} \begin{bmatrix} \mathbf{I}_N \\ \mathbf{P} \end{bmatrix} \right)^{-1} \begin{bmatrix} \mathbf{I}_N \, \mathbf{P}' \end{bmatrix} \begin{bmatrix} (\tau\Sigma)^{-1} & \\ & \Omega^{-1} \end{bmatrix} \begin{bmatrix} \Pi \\ \mathbf{q} \end{bmatrix} \\
&= \left(\begin{bmatrix} \mathbf{I}_N \, \mathbf{P}' \end{bmatrix} \begin{bmatrix} (\tau\Sigma)^{-1} \\ \Omega^{-1}\mathbf{P} \end{bmatrix} \right)^{-1} \begin{bmatrix} \mathbf{I}_N \, \mathbf{P}' \end{bmatrix} \begin{bmatrix} (\tau\Sigma)^{-1}\Pi \\ \Omega^{-1}\mathbf{q} \end{bmatrix} \\
&= \left[(\tau\Sigma)^{-1} + \mathbf{P}'\Omega^{-1}\mathbf{P} \right]^{-1} \left[(\tau\Sigma)^{-1}\Pi + \mathbf{P}'\Omega^{-1}\mathbf{q} \right].
\end{aligned}
$$

This estimator is then used with the standard mean-variance problem formulation, e.g. eq. (2.2) or eq. (2.7). Practical experience with this model, documenting the much greater stability of the resulting portfolio weights than would otherwise be obtained, is related in Bevan and Winkelmann (1998), Litterman (2003), and Fabozzi *et al.* (2006).

2.8.4 Portfolio Resampling

The Black-Litterman estimator still operates before portfolio optimization takes place; its benefits can be traced to a reduced "impedance mismatch" between the expected return estimator and the associated covariance matrix. In contrast, portfolio resampling techniques (Michaud, 1998; Scherer, 2002) attempt to make direct use of the forecast distribution of returns by repeatedly drawing a large number of (*expected-return*, *covariance-matrix*) pairs, and for each computing an efficient frontier, namely a set of (*portfolio-return*, *portfolio-risk*) pairs, over some reasonable risk range. Then those efficient frontiers are averaged over all drawings, and the resulting frontier used to make an allocation decision. Markowitz and Usmen (2003) compare this approach to one similar to the Bayesian approach of p. 31 and observe a good performance of the resampling approach.

A practical limitation to the approach is with respect to *portfolio constraints*: in general, there is no guarantee that the averaged portfolio weights (after resampling) will obey the inequality constraints set in the original optimization problem. Also, due to the high number of optimization steps it requires, it is computationally expensive.

2.8.5 Robust Portfolio Allocation

In recent years, several reformulations of the mean-variance problem have received wide attention that attempt to incorporate estimation uncertainty within the optimization step—not "before", as for the Black-Litterman model, or "around" as for portfolio resampling. They are collectively known as *robust optimization* tech-

niques, and are related to minimax estimators in decision theory.[35] Robust methods
in mathematical programming were introduced by Ben-Tal and Nemirovski (1999)
and further studied in a portfolio choice context by Goldfarb and Iyengar (2003)
and Tütüncü and Koenig (2004) among others. Fabozzi *et al.* (2007) provides a
good survey of the current literature.

The starting point of these approaches is to consider the *uncertainty set* of the
model parameters (the next-period expected returns and their covariances for a portfolio problem) and to ask: "what is the worst-case realization of model parameters
that can arise?", and from there to maximize the utility of this worst-case outcome.
Consider the simplest type of uncertainty region given in the form of "box" intervals

$$\mathcal{U} = \{(\mu, \Sigma) : \mu_L < \mu < \mu_U, \Sigma_L < \Sigma < \Sigma_U, \Sigma \succ 0\},$$

where in this context the $<$ operator should be interpreted elementwise for both
vectors and matrices.

The robust portfolio problem with quadratic utility is expressed as

$$\max_{\mathbf{w}} \left\{ \min_{(\mu, \Sigma) \in \mathcal{U}} \mu' \mathbf{w} - \lambda \mathbf{w}' \Sigma \mathbf{w} \right\}$$

which for the above form of the uncertainty region separates out as

$$\max_{\mathbf{w}} \left\{ \min_{\mu \in \mathcal{U}^\mu} \mu' \mathbf{w} + \max_{\Sigma \in \mathcal{U}^\Sigma} \lambda \mathbf{w}' \Sigma \mathbf{w} \right\}.$$

This can be expressed as a saddle-point problem and solved in polynomial time
(Halldórsson and Tütüncü, 2003). Simpler results can be obtained by considering
other types of uncertainty sets; for instance, when only uncertainty in expected returns is considered, the box constraints reduce to a quadratic program of nearly
the same complexity as the original mean-variance problem; similarly, an ellip-
soidal constraint set yields a second-order cone program (SOCP), which is efficiently
solved by interior-point methods (Boyd and Vandenberghe, 2004). More recently,
Bertsimas and Pachamanova (2008) studied a number of robust optimization approaches to the multiperiod portfolio problem (see next section) in the presence of
transaction costs; in particular, they advocate linear formulations that yield significant computational savings.

2.8.6 *Portfolio Robustness: a Synthesis?*

In light of the large variety of proposed methods for improving the performance
of mean-variance allocation, one may wonder if a particular method turns out to
be "best". To the author's knowledge, a systematic comparison between all of the

---

[35] Robust optimization should not be confused with *robust estimation* in statistics, devoted to
establishing the properties of outlier-resistant estimators.

approaches presented in this section has yet to be published. However, an element of insight has recently been provided by DeMiguel *et al.* (2009), who compare 14 different models on a number of datasets (including U.S. and world equity markets) on three criteria: the out-of-sample Sharpe ratio, certainty equivalent return (from the perspective of a mean-variance investor) and portfolio turnover. On these measures, it is found that *none of the "sophisticated" models consistently beat the naïve* $1/N$ *benchmark* (uniform portfolio weights), *out of sample*. These results suggest that, for the models considered, estimation error still largely dominates any gains obtained from "optimal" diversification.

Chapter 3
Multiperiod Problems

> *Life can only be understood backwards; but it must be lived forwards.*
>
> — *Søren Kierkegaard*

MULTIPERIOD PROBLEMS consider the more general case where an investor makes a *sequence* of decisions, each possibly impacting the following ones. The objective is to find, at each period, the allocation decision that take into consideration a future changing opportunity set (i.e. availability of assets and their risk–return characteristics), the remaining investment horizon, eventual transaction costs, and other constraints such as the desire for intermediate consumption, minimization of tax impact, or the influx of additional capital due to labor income. These decisions, in general, are not identical to those obtained under the myopic (one-period) case, although they can be under specific assumptions (see §3.3/p. 48); more often, we shall see that the optimal solution is constructed from the myopic one as a starting point which is perturbed by a *hedging demands* term to account for "the future". This term makes the obtained portfolio policies differ from iterated single-period ones.

Although Markowitz (1959) discussed the use of dynamic programming to solve the sequential optimal portfolio choice problem (using a time-separable log-utility of consumption as the objective function), he disregarded its more systematic application as computationally unfeasible:

> "For the actual choice of portfolio, however, the dynamic programming techniques cannot be used. They require too much both from man and machine: 1. From the investor they require a utility function $U(C_1, C_2, \ldots, C_t)$. [...] It is no small task to derive a reasonable single period utility function. [...] To attempt to derive a representative utility function for consumption over time, if feasible at all, is nothing short of a major research project. 2. Even with the simplest of utility functions, the requirements for the dynamic programming computation are far beyond economic justification." (p. 278)

Just as the single-period problem, the multiperiod generalization has a rich history, albeit a more academic one.[1] Samuelson (1969) and Merton (1969) are generally credited with posing the general multiperiod consumption and investment problem, Samuelson in discrete time (§3.1/p. 38) and Merton in continuous time (§3.2/p. 45), although Mossin (1968) had previously studied the multiperiod portfolio choice problem (without a consumption aspect). Earlier closely related work includes Tobin (1965) and Phelps (1962) who considers the lifetime utility associated with a consumption history.

After introducing these foundations, we review classical results on the structure of optimal policies (§3.3/p. 48) including a discussion of the optimality conditions of the myopic policy. They are seen to be strongly impacted by the expected evolution of the *investment opportunity set*, namely the risk-reward characteristics of available assets. We give passing mention of an elegant alternative to dynamic programming based on the martingale formulation (§3.4/p. 49) and models that explicitly incorporate consideration of investor learning behavior (§3.5/p. 55). We end this section by giving pointers to common extensions (§3.6/p. 56) that have been proposed.

3.1 The Discrete-Time Case

Consider the problem where at each time-step $t = 0, 1, 2, \ldots, T - 1$ the investor makes a portfolio choice $\mathbf{w}_t$ wherein he tries to intertemporally maximize the expected utility of wealth at the final time T, $U(W(T))$, given a current wealth $W_t \in \mathbb{R}$,

$$\max_{\mathbf{w}_0, \mathbf{w}_1, \ldots, \mathbf{w}_{T-1}} \mathbb{E}_t \left[U(W(T)) \right], \qquad (3.1)$$

subject the the *budget constraint*

$$W_{t+1} = W_t(1 + \mathbf{w}_t' \mathbf{R}_{t+1} + (1 - \mathbf{w}_t' \iota) R_{f,t}), \qquad W_0 \text{ given.} \qquad (3.2)$$

This constraint describes the dynamics of wealth, specifying that the total relative return experienced during period $t + 1$ arises from the allocation $\mathbf{w}_t$ to risky assets and the remainder $(1 - \mathbf{w}_t' \iota)$ from the risk-free asset; note that the latter quantity can be negative, in which case the investor borrows at the risk-free rate.[2] We also require wealth to be always nonnegative, $W_t \geq 0$. Given a sequence of decisions $\{\mathbf{w}_\tau\}_{\tau=t}^{T-1}$, it is useful to observe that the terminal wealth W_T can be written as a function of current wealth W_t,

---

[1] To the author's knowledge, multiperiod optimization has yet to be used in the day-to-day management of an institutional portfolio. This, perhaps, can be attributed to the perceived small gains of the approach compared to its complexity and the remaining inevitable overall portfolio risk.

[2] A more complex constraint can account for differing lending and borrowing rates.

$$W_T = W_t \prod_{\tau=t}^{T-1} \left(1 + \mathbf{w}_\tau' \mathbf{R}_{\tau+1} + (1 - \mathbf{w}_\tau' \iota) R_{f,\tau} \right). \tag{3.3}$$

Consistent with a formulation by dynamic programming (Bellman, 1957; Bertsekas, 2005), it is convenient to express the expected terminal wealth in terms of a *value function*, varying according to the current time,[3] current wealth W_t and other state variables $\mathbf{z}_t \in \mathbb{R}^K, K < \infty$,

$$V(t, W_t, \mathbf{z}_t) = \max_{\{\mathbf{w}_u\}_{u=t}^{T-1}} \mathbb{E}_t \left[U(W_T) \right]$$

$$= \max_{\mathbf{w}_t} \mathbb{E}_t \left[\max_{\{\mathbf{w}_u\}_{u=t+1}^{T-1}} \mathbb{E}_{t+1} \left[U(W_T) \right] \right] \tag{3.4}$$

$$= \max_{\mathbf{w}_t} \mathbb{E}_t \left[V(t+1, W_{t+1}, \mathbf{z}_{t+1}) \right], \tag{3.5}$$

subject to the budget constraint (3.2) and the recursive base case

$$V(T, W_T, \mathbf{z}_T) = U(W_T).$$

The expectations at time t, above, are taken with respect to the joint distribution of asset returns and next state, conditional on the information available at time t, $P(\mathbf{R}_{t+1}, \mathbf{z}_{t+1} | \mathscr{F}_t)$. For our purposes, it shall be sufficient to assume a first-order Markov process for this, such that

$$P(\mathbf{R}_{t+1}, \mathbf{z}_{t+1} | \mathscr{F}_t) = P(\mathbf{R}_{t+1}, \mathbf{z}_{t+1} | \mathbf{R}_t, \mathbf{z}_t);$$

this assumption is not overly restrictive in practice since $\mathbf{z}_t$ can contain (a finite number of) lagged values of relevant variables.

In what follows, we shall use the notation $f_i(\cdot)$ to denote the partial derivative of function f with respect to the i-th argument, e.g.

$$V_2(t', W', \mathbf{z}') \triangleq \left. \frac{\partial V}{\partial W} \right|_{t=t', W=W', \mathbf{z}=\mathbf{z}'}.$$

From eq. (3.5), the first-order conditions for optimality at each time t are obtained as

$$0 = \mathbb{E} \left[V_2(t+1, W_{t+1}, \mathbf{z}_{t+1}) \frac{\partial W_{t+1}}{\partial \mathbf{w}_t} \right]$$

$$= \mathbb{E} \left[V_2 \left(t+1, W_t (1 + \mathbf{w}_t' \mathbf{R}_{t+1} + (1 - \mathbf{w}_t' \iota) R_{f,t}), \mathbf{z}_{t+1} \right) \mathbf{R}_{t+1} \right], \tag{3.6}$$

---

[3] Regarding notation, many treatments of finite-horizon discrete-time dynamic programming (e.g. Bertsekas 2005) simply consider a set of value functions indexed by the current time-step, V_t; here we specifically include time as an explicit variable to preserve notational consistency with the continuous-time treatment in §3.2/p. 45.

These optimality conditions assume that the state variable $\mathbf{z}_{t+1}$ is not impacted by the decision $\mathbf{w}_t$.[4] The second-order conditions are satisfied if the utility function is concave.

Mossin (1968) studied this problem under the assumption of independence of returns across time-steps, no transaction costs and no intermediate consumption. He derived conditions for which the myopic policy can be optimal (§3.3/p. 48). Samuelson (1969) studied the related problem in which the investor derives utility from intermediate consumption and tries to maximize both the discounted utility of the consumption stream and the utility of terminal ("bequeathed") wealth.[5]

3.1.1 Power Utility

In general, (3.6) can only be solved numerically. However, some analytic progress can be achieved in the case of the *power utility*,

$$U(W) = \begin{cases} \frac{W^{1-\alpha}}{1-\alpha}, & \alpha \neq 1 \\ \ln W, & \text{otherwise,} \end{cases}$$

where α is a coefficient of relative risk aversion. This is an example of a constant relative risk aversion (CRRA) utility function, discussed in §2.4/p. 11. In this case (assuming, for simplicity, $\alpha \neq 1$), substituting in eq. (3.4), we obtain

$$V(t, W_t, \mathbf{z}_t) = \max_{\mathbf{w}_t} \mathbb{E}_t \left[\max_{\{\mathbf{w}_\tau\}_{\tau=t+1}^{T-1}} \mathbb{E}_{t+1} \left[\frac{W_T^{1-\alpha}}{1-\alpha} \right] \right]$$

$$= \max_{\mathbf{w}_t} \mathbb{E}_t \left[\max_{\{\mathbf{w}_\tau\}_{\tau=t+1}^{T-1}} \mathbb{E}_{t+1} \left[\frac{\left(W_t \prod_{\tau=t}^{T-1} \left(1 + \mathbf{w}_\tau' \mathbf{R}_{\tau+1} + (1 - \mathbf{w}_\tau' \imath) R_{f,\tau}\right) \right)^{1-\alpha}}{1-\alpha} \right] \right]$$

$$= \max_{\mathbf{w}_t} \mathbb{E}_t \left[\underbrace{\frac{\left(W_t \left(1 + \mathbf{w}_t' \mathbf{R}_{t+1} + (1 - \mathbf{w}_t' \imath) R_{f,t}\right) \right)^{1-\alpha}}{1-\alpha}}_{U(W_{t+1})} \times \right.$$

$$\left. \underbrace{\max_{\{\mathbf{w}_\tau\}_{\tau=t+1}^{T-1}} \mathbb{E}_{t+1} \left[\left(\prod_{\tau=t+1}^{T-1} \left(1 + \mathbf{w}_\tau' \mathbf{R}_{\tau+1} + (1 - \mathbf{w}_\tau' \imath) R_{f,\tau}\right) \right)^{1-\alpha} \right]}_{\psi(t+1, \mathbf{z}_{t+1})} \right].$$

---

[4] This would disregard, for instance, the market impact of trading for large market players. Kissell and Glantz (2003) consider market impact at length.

[5] Samuelson imposes the "greedy granny" condition, i.e. a zero-bequest requirement as a boundary condition.

In the next-to-last expression, the specific form of the power utility allows W_t to be factored out of the maximizations since it is not impacted by the decision variables $\{\mathbf{w}_u\}_{u=t+1}^{T-1}$. Hence, the last expression shows that the value function factors out into two parts: a first one, that depends on future wealth, and equal to the utility of next-time-step wealth, $U(W_{t+1})$, and a second one that only depends on remaining time horizon and future state variables $\mathbf{z}_{t+1}$, but not future wealth. This can further be reduced by writing

$$V(t, W_t, \mathbf{z}_t) = \frac{(W_t)^{1-\alpha}}{1-\alpha} \psi(t, \mathbf{z}_t)$$

where $\psi(t, \mathbf{z}_t)$ satisfies Bellman's equation in a smaller state space,

$$\psi(t, \mathbf{z}_t) = \max_{\mathbf{w}_t} \mathbb{E}_t \left[\left(1 + \mathbf{w}_t' \mathbf{R}_{t+1} + (1 - \mathbf{w}_t' \iota) R_{f,t}\right)^{1-\alpha} \psi(t+1, \mathbf{z}_{t+1}) \right], \quad (3.7)$$

with recursive base case

$$\psi(T, \cdot) = 1. \tag{3.8}$$

If the returns are Independent and Identically Distributed (IID), the above joint expectation between returns and state variables splits out as

$$\psi(t, \mathbf{z}_t) = \max_{\mathbf{w}_t} \left\{ \mathbb{E}_t \left[\left(1 + \mathbf{w}_t' \mathbf{R}_{t+1} + (1 - \mathbf{w}_t' \iota) R_{f,t}\right)^{1-\alpha} \right] \right\} \mathbb{E}_t \left[\psi(t+1, \mathbf{z}_{t+1}) \right], \quad (3.9)$$

where it is readily seen that the optimal portfolio weights at each time-step are independent of the state variables and remaining time horizon, thence must be constant. Put differently, for IID returns (and power utility), there is no difference between the dynamic and myopic portfolios; this property is revisited in §3.3/p. 48.

3.1.2 Numerical Example

Given a model of the conditional return distribution, the Bellman equations (3.7)–(3.8) can be solved numerically. For a power-utility investor, on a two-asset problem—shifting wealth between a riskless bond and a single risky asset—and using the generative model of eq. (2.16), conditioning excess return on dividend yield, some instructive results appear in Fig. 3.1.[6]

The left panel shows the fraction of wealth invested in the risky asset as a function of the initial dividend yield—in effect at the time of making the forecast—and various investor risk aversion levels, for the single-period problem (horizon=1). The

---

[6] The simulations are carried out by estimating the expectation in (3.7) by Monte Carlo sampling with 2500 trajectories. Maximization is performed by numerical optimization using Mathematica 6's built-in NMaximize function for constrained maximization without necessitating the availability of gradients.

plot clearly shows that lower-aversion individuals shift their allocation very rapidly for increasing forecasted returns (as indicated by the dividend yield) in the risky asset: this corresponds to an increased propensity for *market timing* as risk aversion decreases.

The right panel illustrates the *horizon effects* that arise in the presence of return forecastability (but are, as noted above, absent when returns are assumed IID). It shows the allocation to the risky asset as a function of investment horizon, for various *initial* (i.e. first-period) dividend yields and a constant risk aversion $\alpha = 5$. No matter how bleak the immediate prospects for risky returns are (i.e. low dividend yield), a long-horizon investor allocates more to the risky asset than a short-horizon one, since the returns will eventually revert to their unconditional mean over a long period; however, in the model of eq. (3.7), this mean-reversion takes time due to the high autocorrelation in the (log) dividend yield.

A more complete picture of the optimal policy, and associated value function, is given in Fig. 3.2.

3.1.3 The Mean-Variance Multiperiod Criterion

Surprisingly, it has been only relatively recently that a multiperiod analog to the mean-variance problem received a thorough solution in the discrete-time case.[7] Li and Ng (2000) analyzed various formulations of the maximization of terminal quadratic utility under several hypotheses, provided explicit solutions in simplifying cases, and derived analytical expressions for the multiperiod mean-variance efficient frontier (a concept that had, until that point, received no attention in the multiperiod case).

More specifically, considering the N-risky-asset case as previously, the form of the mean-variance optimization problem follows the minimum-variance formulation (2.2)–(2.4) or the utility-maximization formulation (2.7)–(2.8), with the exception that the objective function is expressed in terms of terminal *wealth* instead of portfolio relative return. For instance, the utility formulation takes the form

$$\max_{\{\mathbf{w}_t\}_{t=1}^{T-1}} \quad \mathbb{E}[W_T] - \lambda \operatorname{Var}[W_T] \qquad (3.10)$$

$$\text{subject to} \quad W_{t+1} = W_t \mathbf{w}_t'(1 + \mathbf{R}_{t+1}) \qquad (3.11)$$

$$\iota'\mathbf{w} = 1, \qquad (3.12)$$

where the initial wealth W_0 is given. The derivation of optimal solutions is complicated by the fact that the objective is not time-separable (in the dynamic programming sense), but analytical solutions exist if asset returns are assumed independent between periods; they do not need to be identically distributed, provided that all future return means and covariance matrices are known ahead of time. Obviously,

---

[7] In continuous time, the problem was solved by Korn and Trautmann (1995) and Zhou (2000).

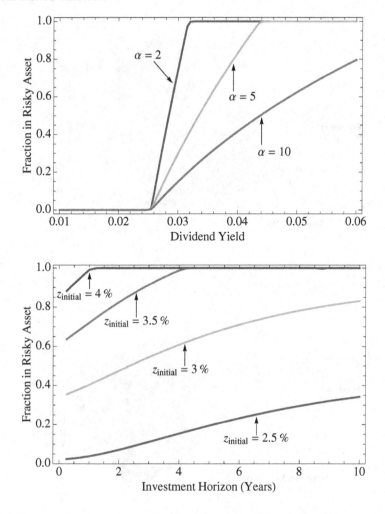

Fig. 3.1 Top: Fraction of wealth invested in the risky asset for the two-asset problem as a function of the initial dividend yield, for an investment horizon of one period (one quarter, in this case) and various investor risk aversion levels (α). **Bottom:** Fraction of wealth invested in the risky asset as a function of the investment horizon, for various initial dividend yields and constant risk aversion $\alpha = 5$.

the estimation methodology of §2.7/p. 23 can be put to bear for this task. Moreover, §3.5/p. 55 connect these mildly unrealistic assumptions with the fact that in a multiperiod setting, the optimal policy depends on the fact that we *expect* to learn more about the asset return distribution in the future.

Leippold *et al.* (2004) provided an interpretation of the solution to the multiperiod mean-variance problem in terms of an orthogonal set of basis strategies, each with a clear economic interpretation. They use this analysis to provide analytical

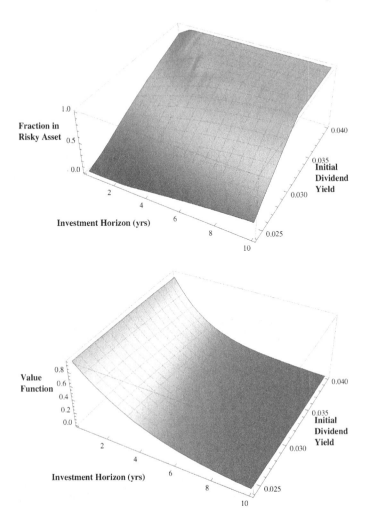

Fig. 3.2 Top: Optimal policy (fraction of capital invested in the risky asset) as a function of time-to-maturity (years) and initial dividend yield, for an investor with a constant risk-aversion $\alpha = 5$. **Bottom:** Value function under the same conditions.

solutions to portfolios consisting of both assets and liabilities. More recently, Cvitanić *et al.* (2008) connect this problem to a specific case of multiperiod Sharpe ratio maximization.

3.2 The Continuous-Time Case

The continuous-time analysis is due to Merton (1969; 1971) and illustrates the analytical tractability of the approach; Merton's seminal papers consider the joint optimization of investment and consumption decisions, including a number of variations such as the effect of wage income and alternative stochastic processes. Our succinct exposition draws from Brandt (2004) and for simplicity, only considers the maximization of the terminal utility of wealth—disregarding intermediate consumption—and assumes that all assets are driven by log-diffusion processes (geometric Brownian motion) and can be traded continuously without friction (transaction costs, taxes) or consideration of background risk (general economic downturn, unemployment risk).[8] Despite this simplified setting, the results obtained are sufficiently illuminating to convey noteworthy intuition about the structure of the optimal multiperiod portfolio choice.

In continuous time, the problem formulation is identical to the discrete-time objective (3.1), except that instead of making a discrete set of decisions, a continuous allocation trajectory must be found, subject to a continuous-time budget constraint. We shall assume, for $0 \leq t < T, t \in \mathbb{R}$, that the N risky asset prices $\mathbf{P}_t$ and K-dimensional state vector $\mathbf{z}_t$ evolve jointly according to correlated Itô vector processes,[9]

$$\frac{d\mathbf{P}_t}{\mathbf{P}_t} = (\mu^{\mathbf{P}}(\mathbf{z}_t,t) + r_f)dt + \mathbf{D}^{\mathbf{P}}(\mathbf{z}_t,t)d\mathbf{B}_t^{\mathbf{P}} \qquad (3.13)$$

$$d\mathbf{z}_t = \mu^{\mathbf{z}}(\mathbf{z}_t,t)dt + \mathbf{D}^{\mathbf{z}}(\mathbf{z}_t,t)d\mathbf{B}_t^{\mathbf{z}} \qquad (3.14)$$

subject to the budget constraint

$$\frac{dW_t}{W_t} = (\mathbf{w}_t'\mu_t^{\mathbf{P}} + r_f)dt + \mathbf{w}_t'\mathbf{D}_t^{\mathbf{P}}d\mathbf{B}_t^{\mathbf{P}} \qquad (3.15)$$

where $\mu^{\mathbf{P}}(\mathbf{z}_t,t)$ is the conditional mean excess return of the assets, $\mathbf{D}^{\mathbf{P}}(\mathbf{z}_t,t)$ is the conditional $N \times N$ price process diffusion matrix, $\mu^{\mathbf{z}}(\mathbf{z}_t,t)$ is the conditional drift of the state variables, and $\mathbf{D}^{\mathbf{z}}(\mathbf{z}_t,t)$ is the conditional $K \times K$ state process diffusion matrix. For readability, we drop the explicit conditioning on $(\mathbf{z}_t,t)$ in what follows, and use a simple t subscript for the previous quantities. The diffusion matrices $\mathbf{D}_t^{\mathbf{P}}$ and $\mathbf{D}_t^{\mathbf{z}}$ respectively induce covariance matrices $\Sigma_t^{\mathbf{P}}$ and $\Sigma_t^{\mathbf{z}}$ within the price process $\mathbf{P}_t$ and state process $\mathbf{z}_t$. Furthermore, we assume that the underlying Brownian pro-

---

[8] See §3.6/p. 56 for references to the many extensions that have been proposed to address these restrictions. See Merton (1990) and Duffie (2001) for more formal treatments of the material in this section.

[9] This section assumes some familiarity with stochastic differential equations; see §A.3/p. 74 for a review of Itô's lemma, used in the derivations to follow.

cesses $\mathbf{B}_t^{\mathbf{P}}$ and $\mathbf{B}_t^{\mathbf{z}}$ are related by a time-varying $N \times K$ correlation matrix ρ_t.[10] The notation $d\mathbf{P}_t/\mathbf{P}_t$ should be interpreted as elementwise differentiation.

To derive the continuous-time Bellman equation, we can proceed informally as follows (see Merton (1971) for a more complete treatment). We obtain it as the limit as $\Delta t \to 0$ of the discrete-time Bellman equation (3.5). First the equation is rewritten as

$$0 = \max_{\mathbf{w}_t} \mathbb{E}_t \left[V(t+1, W_{t+1}, \mathbf{z}_{t+1}) - V(t, W_t, \mathbf{z}_t) \right]$$

and we replace the transition to the "next" time-step by an interval Δt,

$$0 = \max_{\mathbf{w}_t} \mathbb{E}_t \left[V(t+\Delta t, W_{t+\Delta t}, \mathbf{z}_{t+\Delta t}) - V(t, W_t, \mathbf{z}_t) \right],$$

which yields, in the limit of $\Delta t \to 0$,

$$0 = \max_{\mathbf{w}_t} \mathbb{E}_t \left[dV(t, W_t, \mathbf{z}_t) \right]. \tag{3.16}$$

We can then mechanically apply Itô's lemma (A.20) (see p. 76) to the value function to derive (for notational convenience, V is used in place of $V(t, W_t, \mathbf{z}_t)$ when no confusion is possible)

$$dV = V_1 dt + V_2 dW_t + V_3 d\mathbf{z}_t + \frac{1}{2} V_{2,2} dW_t^2 + V_{2,3} dW_t d\mathbf{z}_t + \frac{1}{2} V_{3,3} d\mathbf{z}_t^2,$$

where, as previously, the notation V_i denotes the partial derivative of V with respect to the i-th argument (note that V_1 and V_2 are scalars, whereas V_3 is a K-vector). Substituting $d\mathbf{z}_t$ and dW_t respectively from eq. (3.14) and (3.15) and applying the usual rules for the product of differentials[11] we obtain

$$dV = V_1 dt + V_2 W_t (\mathbf{w}_t' \boldsymbol{\mu}_t^{\mathbf{P}} + r_f) dt + V_2 W_t \mathbf{w}_t' \mathbf{D}_t^{\mathbf{P}} d\mathbf{B}_t^{\mathbf{P}} +$$
$$V_3' \boldsymbol{\mu}_t^{\mathbf{z}} dt + V_3' \mathbf{D}_t^{\mathbf{z}} d\mathbf{B}_t^{\mathbf{z}} + \frac{1}{2} V_{2,2} W_t^2 \mathbf{w}_t' \mathbf{D}_t^{\mathbf{P}} I_N \mathbf{D}_t^{\mathbf{P}'} \mathbf{w}_t dt +$$
$$W \mathbf{w}_t' \mathbf{D}_t^{\mathbf{P}} \rho_t' \mathbf{D}^{\mathbf{z}'} V_{2,3} dt + \frac{1}{2} \mathrm{tr} \left[\mathbf{D}_t^{\mathbf{z}} I_K \mathbf{D}_t^{\mathbf{z}'} V_{3,3} \right] dt.$$

Taking expectations, substituting back into eq. (3.16), dividing left and right sides by dt, and rearranging terms, we obtain the continuous-time Bellman equation,

---

[10] Note that the elements *within* both $d\mathbf{B}_t^{\mathbf{P}}$ and $d\mathbf{B}_t^{\mathbf{z}}$ are uncorrelated; all the "inner" correlation structure within the processes $\mathbf{P}_t$ and $\mathbf{z}_t$ is induced though the off-diagonal terms in the diffusion matrices $\mathbf{D}_t^{\mathbf{P}}$ and $\mathbf{D}_t^{\mathbf{z}}$.

[11] Namely,

$$(dt)^2 = 0, \qquad\qquad dt\,(d\mathbf{B}_t^{\{\mathbf{P},\mathbf{z}\}})_i = 0,$$
$$(d\mathbf{B}_t^{\mathbf{P}})_i (d\mathbf{B}_t^{\mathbf{P}})_j = \delta_{i,j} dt, \qquad (d\mathbf{B}_t^{\mathbf{z}})_i (d\mathbf{B}_t^{\mathbf{z}})_j = \delta_{i,j} dt, \qquad (d\mathbf{B}_t^{\mathbf{P}})_i (d\mathbf{B}_t^{\mathbf{z}})_j = \rho_{i,j} dt.$$

$$0 = \max_{\mathbf{w}_t}\left[V_1 + W_t(\mathbf{w}_t'\boldsymbol{\mu}_t^{\mathbf{P}} + r_f)V_2 + \boldsymbol{\mu}_t^{\mathbf{z}'}V_3 + \right.$$
$$\left. \frac{1}{2}W_t^2\mathbf{w}_t'\boldsymbol{\Sigma}_t^{\mathbf{P}}\mathbf{w}_tV_{2,2} + W_t\mathbf{w}_t'\mathbf{D}_t^{\mathbf{P}}\boldsymbol{\rho}_t'\mathbf{D}_t^{\mathbf{z}'}V_{2,3} + \frac{1}{2}\mathrm{tr}\left[\boldsymbol{\Sigma}_t^{\mathbf{z}}V_{3,3}\right]\right] \quad (3.17)$$

subject to terminal conditions $V(T,W_T,\mathbf{z}_T) = U(W_T)$. The first-order conditions for optimality are obtained as a stationary point of eq. (3.17) with respect to $\mathbf{w}_t$,

$$\boldsymbol{\mu}_t^{\mathbf{P}}V_2 + W_tV_{2,2}\boldsymbol{\Sigma}_t^{\mathbf{P}}\mathbf{w}_t + \mathbf{D}_t^{\mathbf{P}}\boldsymbol{\rho}_t'\mathbf{D}_t^{\mathbf{z}'}V_{2,3} = 0,$$

which can explicitly be solved for the optimal portfolio weights

$$\mathbf{w}_t^* = -\frac{V_2}{W_tV_{2,2}}\left(\boldsymbol{\Sigma}_t^{\mathbf{P}}\right)^{-1}\boldsymbol{\mu}_t^{\mathbf{P}} \qquad \left.\right\} \quad \begin{array}{l}\text{Myopic}\\\text{Portfolio}\end{array}$$

$$\qquad -\frac{V_2}{W_tV_{2,2}}\frac{V_{2,3}}{V_2}\left(\boldsymbol{\Sigma}_t^{\mathbf{P}}\right)^{-1}\mathbf{D}_t^{\mathbf{P}}\boldsymbol{\rho}_t'\mathbf{D}_t^{\mathbf{z}'} \quad \left.\right\} \quad \begin{array}{l}\text{Hedging}\\\text{Demands}\end{array}$$

$$(3.18)$$

The "myopic portfolio" term corresponds to the solution of the one-period problem and is equivalent to eq. (2.10). The factor $-V_2/(W_tV_{2,2})$ represents the investor's relative risk tolerance (reciprocal of the relative risk aversion). The "hedging demands" term corresponds to an additional demand for risky assets resulting from changes in the investment opportunity set. It depends on the following factors:

- Non-constant state variables (the $\mathbf{D}_t^{\mathbf{z}}$ matrix must be non-zero);
- Correlation between state variables and risky-asset prices ($\boldsymbol{\rho}_t$);
- How strongly the changes in state variables $\mathbf{z}_t$ affect the utility of wealth ($V_{2,3}$ factor).

If any of those factors is zero, hedging demands disappear and only the myopic portfolio remains. Hence, the presence of hedging demands depends on the ability of state variables to capture (instantaneous) changes in the asset price process. The marginal utility of wealth with respect to the state variables ($V_{2,3}/V_2$) chooses the appropriate trade-off between the myopic and hedging terms. Brandt (2004) offers the following interpretation on the relationship between state variable and asset price processes: "The projection $\left((\boldsymbol{\Sigma}_t^{\mathbf{P}})^{-1}\mathbf{D}_t^{\mathbf{P}}\boldsymbol{\rho}_t'\mathbf{D}_t^{\mathbf{z}'}\right)$ delivers the weights of K portfolios that are maximally correlated with the state variable innovations and the derivatives of marginal utility with respect to the state variables measure how important each of these state variables is to the investor. Intuitively, the investor takes positions in each of the maximally correlated portfolios to partially hedge against undesirable innovations in the state variables."

Merton (1973) presents a very elegant version of the CAPM wherein all investors are assumed to be intertemporal maximizers (as opposed to single-period Markowitz maximizers as in the original CAPM; see §2.7.1/p. 23) and considers equilibrium relations among expected returns; as such, he derives solutions for the price of risk that are quite different from the CAPM's market β. In particular, it is shown that even though a risky asset exhibits no "systematic" (or market) risk, it can earn a return different from the risk-free rate due to the hedging demands introduced above.

3.3 Structure of Optimal Solutions

A major concern in the classical analyses is with respect to the *structure of optimal solutions*. In continuous time, the problem can be solved analytically only for a handful of special cases; for instance, in the case of the power utility, one can assume—just as for discrete-time—a separable solution of the form

$$V(t, W_t, \mathbf{z}_t) = \frac{W_t^{1-\lambda}}{1-\lambda} \psi(t, \mathbf{z}_t), \tag{3.19}$$

which can be substituted in eq. (3.18) to yield optimal portfolio weights, and then in Bellman's equation (3.17) to yield a partial differential equation involving only $\psi(\cdot, \cdot)$,

$$\psi_1 + (1-\lambda)(\mathbf{w}_t^{*\prime}\mu_t^{\mathbf{P}} + r_f)\psi + \mu_t^{\mathbf{z}\prime}\psi_2 - \frac{\lambda(1-\lambda)}{2}\mathbf{w}_t^{*\prime}\Sigma_t^{\mathbf{P}}\mathbf{w}_t^*\psi$$
$$+ (1-\lambda)\mathbf{w}_t^{*\prime}\mathbf{D}_t^{\mathbf{P}}\rho_t'\mathbf{D}_t^{\mathbf{z}\prime}\psi_2 + \frac{1}{2}\mathrm{tr}\left[\Sigma_t^{\mathbf{z}}\psi_{2,2}\right] = 0, \tag{3.20}$$

with boundary condition $\psi(T, \cdot) = 1$.

Merton (1971) derived explicit solutions for the Hyperbolic Absolute Risk Aversion (HARA) family of utility functions (see §3.6.2/p. 56), which encompass the power utility case. In particular he showed that, for log-normally distributed assets and solving the case of the more general optimal consumption and investment problem,[12] that optimal consumption and investment policies have a form linear in current wealth,

$$C_t^* = a(t)W_t + b(t), \qquad\qquad \mathbf{w}_t^* W_t = g(t)W_t + h(t),$$

with a, b, g, h at most functions of time, if *and only if* the investor's utility functions on both consumption and terminal wealth belongs to the HARA family.

In more recent work, Kim and Omberg (1996) derive closed-form solutions for a number of specific parametrizations of the HARA utility in the case of a constant risk-free rate and a single risky asset with a stochastic risk premium following an Ornstein–Uhlenbeck (mean-reverting) process.[13]

---

[12] See §3.6/p. 56.

[13] The risk premium is the excess return (over the risk-free rate) paid by the market for enticing investors to hold a risky asset.

3.3.1 Logarithmic Utility

Just as for the single-period problem, the logarithmic utility brings useful simplification in the multiperiod case. Logarithmic utility is a limiting form of the power utility where $\lambda = 1$. Substituting in eq. (3.20) yields the simplified equation

$$\psi_1 + \mu_t^{z'} \psi_2 + \frac{1}{2} \text{tr} \left[\Sigma_t^z \psi_{2,2} \right] = 0,$$

subject to the boundary condition $\psi(T, \cdot) = 1$. It is obvious that the constant function $\psi(\cdot, \cdot) \equiv 1$ is a solution. Substituting in eq. (3.18), we see that the hedging demands term disappears (since $V_{2,3} \equiv 0$), leaving as the optimal solution only the myopic portfolio

$$\mathbf{w}_t^* = \left(\Sigma_t^P \right)^{-1} \mu_t^P.$$

A similar result also obtains in the discrete-time case. We now review the conditions under which the optimal multiperiod choice is in fact the myopic portfolio.

3.3.2 When is the Myopic Policy Optimal?

Mossin (1968) examines the conditions for which the *myopic* portfolio choice is optimal.[14] From the results of eq. (3.9), eq. (3.18) and the previous section, we can summarize the conditions for optimality of the myopic policy as follows:

- The investment opportunities are fixed (for example, in the case of IID returns).
- The investment opportunities vary with time, but are unhedgeable; in this case, the ρ_t correlation matrix between state variables and asset returns is zero in eq. (3.18), and the induced hedging demands are also zero.
- The investor has a logarithmic utility on terminal wealth, as shown in the previous subsection.

3.4 The Martingale Formulation

The martingale formulation for optimal portfolios was introduced by Pliska (1986), Karatzas *et al.* (1987) and Cox and Huang (1989; 1991), and relies on a methodology established by Harrison and Kreps (1979) in the context of contingent claim valuation. Quite remarkably, this approach admits nearly the same class of problems as the optimal control formulation of §3.2/p. 45, yet dispenses with the arduous nonlinear Bellman partial differential equation (3.17), requiring solution only to a static

---

[14] Recall that the myopic choice at time t depends only on the investment opportunity set and investor wealth at that time, disregarding future opportunities completely; in discrete time, it is equivalent to optimizing over the last period in the horizon.

optimization problem (and auxiliary sub-problems, all much easier than eq. (3.17)). In a sense, it transforms an optimal control problem into a far simpler constrained optimization problem.

The approach is reviewed in some detail by Merton (1990, Chapter 6) and contrasted to the dynamic programming formulation. Because of its astuteness, we take time to outline the main ideas of the method, drawing from Merton's presentation but focusing on maximizing the utility of terminal wealth, omitting intermediate consumption.

3.4.1 The Growth-Optimum Portfolio

Let W_t be the value at time t of a portfolio that reinvests all earnings. Let $\mathrm{ACCR}(t,T)$ be the *average continuously compounded return* of the portfolio between times t and T,

$$\mathrm{ACCR}(t,T) \triangleq \frac{1}{T-t} \log\left[\frac{W_T}{W_t}\right].$$

Consider the trading strategy that maximizes this quantity; the resulting portfolio is called a *growth-optimum* portfolio. It is easy to see that it arises from having a log-utility on terminal wealth, $U(W_T) = \log W_T$, since

$$\mathbb{E}_t\left[\mathrm{ACCR}(t,T)\right] = \frac{\mathbb{E}_t\left[\log W_T - \log W_t\right]}{T-t}$$
$$\propto \mathbb{E}_t\left[\log W_T\right] - \log W_t$$

where W_t is known at time t and of no impact in the optimal strategy. From the results of §3.3/p. 48, we have that the optimal portfolio weights in the risky assets, $\mathbf{w}_t^g$, can be written as

$$\mathbf{w}_t^g = (\Sigma_t^{\mathbf{P}})^{-1}\mu_t^{\mathbf{P}} \tag{3.21}$$

where $\Sigma_t^{\mathbf{P}}$ is the instantaneous covariance matrix between asset returns at time t and $\mu_t^{\mathbf{P}}$ is the vector of instantaneous expected excess asset returns. The fraction allocated to the risk-free asset is $1 - \sum_{i=1}^{N}(\mathbf{w}_t^g)_i$.[15] As established previously, the growth-optimal portfolio rule is myopic.

Let X_t be the value of the growth-optimum portfolio at time t. From the posited asset-price dynamics of eq. (3.13), the dynamics of X_t are given as

$$dX_t = \left[\mathbf{w}_t^{g\prime}\left[\frac{d\mathbf{P}_t}{\mathbf{P}_t} - r_f\,dt\right] + r_f\,dt\right]X_t$$
$$= (\bar{\mu}^2 + r_f)X_t\,dt + \bar{\mu}X_t\,dz, \tag{3.22}$$

---

[15] As always, a negative fraction corresponds to borrowing at the risk-free rate.

where $\bar{\mu}^2 \triangleq \mu_t^{\mathbf{P}\prime}(\Sigma_t^{\mathbf{P}})^{-1}\mu_t^{\mathbf{P}}$ and $\mathrm{d}z \triangleq \mu_t^{\mathbf{P}\prime}(\Sigma_t^{\mathbf{P}})^{-1}\mathbf{D}_t^{\mathbf{P}}\mathrm{d}\mathbf{B}_t^{\mathbf{P}}/\bar{\mu}$ follow from eq. (3.21). The increment $\mathrm{d}z$ is a standard Wiener process.

3.4.2 The Cox-Huang Method

In continuous-time, if there is no intervening consumption, the optimal portfolio choice problem can be formulated as

$$\max_{\{\mathbf{w}_t\}_{t=0}^T} \mathbb{E}_0\big[U(W_T)\big], \tag{3.23}$$

subject to the budget constraint (3.15) and feasibility restriction $W_t \geq 0$ for all $t \leq T$. The dynamic programming solution studied previously expresses the optimal solution in a "feedback control" form, wherein the action depends on the current state of the process being controlled, $\mathbf{w}_t^* = \mathbf{w}^*(t, W_t, \mathbf{P}_t)$ (omitting other state variables $\mathbf{z}_t$ for simplicity). The expectation at time 0 is taken with respect to the joint distribution of asset prices $\mathbf{P}_t$ and current wealth dynamics W_t.

Now consider a different problem specified as

$$\max_{\{\mathbf{w}_t\}_{t=0}^T} \tilde{\mathbb{E}}_0\big[U(\hat{W}_T)\big] \tag{3.24}$$

$$\text{subject to} \quad \hat{W}_T \geq 0,$$

$$X_0\tilde{\mathbb{E}}_0\Big[\frac{\hat{W}_T}{X_T}\Big] \leq W_0, \tag{3.25}$$

where X_t is the value of the growth-optimum portfolio at time t, as defined previously. The expectation $\tilde{\mathbb{E}}$ is taken with respect to the joint distribution of asset prices $\mathbf{P}_t$ and values of the growth-optimum portfolio X_t. In particular, it does not include consideration of current wealth or other aspects of the controlled process. This implies that the optimal terminal wealth function $\hat{W}_T^* \equiv H(T, \hat{X}_T, \mathbf{P}_T; X_0, W_0, \mathbf{P}_0)$ *will not* have a "feedback control" form, since portfolio choices $\{\mathbf{w}_t\}_{t=0}^T$ and thence the wealth trajectory $\hat{W}_t$ have no bearing on the asset price process $\mathbf{P}_t$ or value of the growth-optimum portfolio X_t.[16] Put differently, eq. (3.24) can be solved by using the Karush-Kuhn-Tucker (KKT) conditions for *static* constrained optimization.

The connection between problems (3.23) and (3.24) is established by the following result by Cox and Huang (1991):

Theorem 1 (Cox–Huang Equivalence). *Under quite mild regularity conditions, there exists a solution to (3.23) if and only if (a) there exists a solution to (3.24), and (b) $W_T = \hat{W}_T^*$.*

In-depth economic intuition behind this result is provided by Merton (1990, Chapter 16). In the remainder of this section, we derive the optimal portfolio rule $\mathbf{w}_t^*$ that

---

[16] Assuming negligible market impact.

arises from solution to eq. (3.24). We start by incorporating the constraints into the objective by way of Lagrange multipliers,

$$\max_{\{\mathbf{w}_t\}_{t=0}^{T}, \lambda_1, \lambda_2} \tilde{\mathbb{E}}_0 \left[U(\hat{W}_T) + \lambda_1 \left[W_0 - \frac{X_0 \hat{W}_T}{X_T} \right] + \lambda_2 \hat{W}_T \right], \qquad (3.26)$$

where $\lambda_1, \lambda_2 \geq 0$. For all X_t and $\mathbf{P}_t$ (having positive probability in the above expectation), the first-order condition for optimality is written as

$$U_1(\hat{W}_T) = \lambda_1 \frac{X_0}{X_T} - \lambda_2. \qquad (3.27)$$

The substantive analysis can proceed assuming that constraint (3.25) is binding with equality, for otherwise the investor's initial wealth is sufficient to ensure satiation given his utility function, and the optimal policy is therefore to invest in the riskless asset.[17] The assumption of non-satiation ensures that for any terminal wealth W_T, we have strictly positive marginal utility $U_1(W_T) > 0$ and strict concavity of utility, $U_{1,1}(W_T) < 0$. Non-satiation in turn implies that the shadow price of wealth, λ_1, is strictly positive in eq. (3.26).

From the KKT condition[18] $\lambda_2 \hat{W}_T^* = 0$ and eq. (3.27), we have

$$\lambda_2 = \max \left[0, \lambda_1 \frac{X_0}{X_T} - U_1(0) \right]. \qquad (3.28)$$

Furthermore, since $U_{1,1}(\cdot) < 0$, U_1 is invertible. Let $R(y) \triangleq U_1^{-1}(y)$. From eq. (3.27) and (3.28), we determine the optimal terminal wealth to be

$$\begin{aligned}
\hat{W}_T^* &= R \left[\lambda_1 \frac{X_0}{X_T} - \max \left[0, \lambda_1 \frac{X_0}{X_T} - U_1(0) \right] \right] \\
&= R \left[\max \left[\lambda_1 \frac{X_0}{X_T}, U_1(0) \right] \right] \\
&= \max \left[R \left[\lambda_1 \frac{X_0}{X_T} \right], R(U_1(0)) \right] \\
&= \max \left[R \left[\lambda_1 \frac{X_0}{X_T} \right], 0 \right], \qquad (3.29)
\end{aligned}$$

taking advantage of the monotonicity of $R(\cdot)$ to exchange R and max in the third step. The solution to $\hat{W}_T^*$ only requires the determination of λ_1. As indicated above, this is possible assuming that constraint (3.25) is binding with equality. Under this condition, substituting eq. (3.29) in (3.25) yields the following transcendental algebraic equation

---

[17] See Merton (1990, p. 174) for a detailed argument.

[18] See, e.g., Luenberger and Ye (2007) for more details on the KKT conditions for constrained optimization.

$$\tilde{\mathbb{E}}_0 \left[\frac{\max\left[R\left[\lambda_1 \frac{X_0}{X_T}\right], 0\right]}{X_T} \right] - \frac{W_0}{X_0} = 0$$

whose solution for λ_1 depends only on the initial conditions and can be expressed as $\lambda_1 = \lambda_1(X_0, \mathbf{P}_0, W_0)$. For compactness, we shall express $\hat{W}_T^*$ as

$$\hat{W}_T^* \equiv H(T, X_T, \mathbf{P}_T), \tag{3.30}$$

noting that the implicit functional dependence of H on X_0, $\mathbf{P}_0$, and W_0 is omitted for simplicity. We can derive the optimal portfolio strategy $\mathbf{w}_t^*$ as follows. Define the function $F(t, X_t, \mathbf{P}_t)$ as

$$F(t, X_t, \mathbf{P}_t) \overset{\triangle}{=} X_t \, \tilde{\mathbb{E}}_t \left[\frac{H(T, X_T, \mathbf{P}_T)}{X_T} \middle| X_t, \mathbf{P}_t \right].$$

From (3.25), we have that $F(0, X_0, \mathbf{P}_0) = W_0$. At an arbitrary time t, assuming the investor acts optimally since time 0, he faces from time t the same problem (3.24)–(3.25) faced from time 0,

$$\max_{\{\mathbf{w}_\tau\}_{\tau=t}^T} \tilde{\mathbb{E}}_t \left[U(\hat{W}_T) \right] \tag{3.31}$$

$$\text{subject to} \quad \hat{W}_T \geq 0,$$

$$X_t \tilde{\mathbb{E}}_t \left[\frac{\tilde{W}_T}{X_T} \right] \leq W_t. \tag{3.32}$$

Because eq. (3.30) is an intertemporal optimum, it must be the case that H is also the rule the investor would follow as a solution to eq. (3.31) and assuming no satiation, constraint (3.32) remains satisfied with equality. By the definition of F and (3.32), we have

$$W_t = F(t, X_t, \mathbf{P}_t). \tag{3.33}$$

From Itô's lemma, asset-price dynamics (3.13) and the dynamics of the growth-optimal portfolio X_t (3.22), we have, after some algebra, that the optimal wealth dynamics are given by

$$dF = \bar{\alpha}F \, dt + F_2 \bar{\mu} X_t \, dz + (F_3 \odot \mathbf{P}_t)' \mathbf{D}_t^{\mathbf{P}} d\mathbf{B}_t^{\mathbf{P}} \tag{3.34}$$

with

$$\bar{\alpha}F \equiv F_1 + (\bar{\mu}^2 + r_f)XF_2 + \frac{1}{2}\bar{\mu}^2 X^2 F_{2,2} +$$

$$(F_3 \odot \mathbf{P}_t)'(\mu_t^{\mathbf{P}} + r_f) + \frac{1}{2}\text{Tr}\left[F_{3,3}\Sigma_t^{\mathbf{P}}\right] + X(F_{2,3} \odot \mathbf{P}_t)'\mu_t^{\mathbf{P}}$$

where the $\odot$ operator signifies element-wise multiplication of vector elements, i.e. $(\mathbf{x} \odot \mathbf{y})_i \overset{\triangle}{=} \mathbf{x}_i \mathbf{y}_i$.

Theorem 2 (Cox–Huang Optimal Weights). *If there exists an optimal solution to problem (3.24), $\hat{W}_T^*$, then for $t \leq T$ the optimal portfolio strategy $\{\mathbf{w}_t^*\}$ that achieves this allocation is given by*

$$\mathbf{w}_t^* W_t = F_2(t, X_t, \mathbf{P}_t) X_t \, \mathbf{w}_t^g + F_3(t, X_t, \mathbf{P}_t) \odot \mathbf{P}_t \tag{3.35}$$

with the balance of the investor's wealth, $1 - \iota' \mathbf{w}_t^*$, in the riskless asset, where $\mathbf{w}_t^g$ is given by eq. (3.21).

Proof. Let $\{\mathbf{w}_t^*\}$ denote the optimal allocations in the risky assets. From eq. (3.15), the dynamics of wealth under optimal allocation are given as

$$dW = \left(\mathbf{w}_t^{*\prime} \mu_t^{\mathbf{P}} + r_f\right) W_t \, dt + W_t \mathbf{w}_t^{*\prime} \mathbf{D}_t^{\mathbf{P}} d\mathbf{B}_t^{\mathbf{P}}. \tag{3.36}$$

But from eq. (3.33), we must have $dW - dF \equiv 0$ for all $t \leq T$, and comparing eq. (3.34) and (3.36) this can be satisfied if and only if

$$\bar{\alpha} F = \left(\mathbf{w}_t^{*\prime} \mu_t^{\mathbf{P}} + r_f\right) W_t \tag{i}$$

and

$$F_2 \bar{\mu} X_t \, dz + (F_3 \odot \mathbf{P}_t)' \mathbf{D}_t^{\mathbf{P}} d\mathbf{B}_t^{\mathbf{P}} = W_t \mathbf{w}_t^{*\prime} \mathbf{D}_t^{\mathbf{P}} d\mathbf{B}_t^{\mathbf{P}}. \tag{ii}$$

Replacing dz by its definition, we simplify the common terms $\mathbf{D}_t^{\mathbf{P}} d\mathbf{B}_t^{\mathbf{P}}$ on both sides of (ii) and obtain the result.   □

By virtue of Theorem 1, the optimal weights found from eq. (3.35) are also those that solve the original problem (3.23).

The only remaining hurdle in applying the method is to obtain the distribution $P(X_t, \mathbf{P}_t | X_0, \mathbf{P}_0), 0 \leq t \leq T$, under which the expectation $\tilde{\mathbb{E}}$ can be evaluated. This is possible by solving a backward Kolmogorov equation (Merton, 1990; Wilmott, 2006), a linear parabolic partial differential equation, itself much easier to solve than the nonlinear Bellman equation.

Cox and Huang (1989) derive explicit solutions, in the presence of nonnegativity constraints on consumption and final wealth, for hyperbolic absolute risk aversion (HARA) utility functions when the asset prices follow a geometric Brownian motion. The nonnegativity constraints cause the optimal policies to no longer be linear in the moments of the return distribution. Wachter (2002) finds closed-form solutions for mean-reverting returns when markets are assumed to be complete.

3.4.3 Implementation

A number of implementations of this method have been presented in the literature. Cvitanić *et al.* (2003) introduce a relatively simple method based on pure Monte Carlo simulation for approximating the expectations required for computing optimal portfolios; their method assumes that asset-price and state variable dynamics are known. A different simulation approach, by Detemple *et al.* (2003), derives explicit components for the hedging demand terms of optimal portfolios using the Malliavin calculus and generalizes earlier results by Ocone and Karatzas (1991); the approach allows a large number of assets and state variables, assumed to follow a diffusion process, to be used. The convergence and efficiency properties of the Malliavin derivatives, in contrast to PDE and other Monte Carlo estimators, are analyzed. In more recent work, Aït-Sahalia and Brandt (2007) use option-market prices to directly infer state prices; they find significantly different optimal consumption and investment policies than those arising from standard assumptions on asset return dynamics.

Of a related flavor is the work by Brandt and Santa-Clara (2006) who consider augmenting the asset space by a set of managed portfolios, both *conditional portfolios* that are proportional to conditioning variables, and *timing portfolios* that invest in one asset for one time period (at some point in the future) and do not invest during other periods. These assets are similar in spirit to the Arrow–Debreu securities which form the theoretical foundation of the Cox–Huang method. Brandt and Santa-Clara show that solving a *static* Markowitz mean-variance problem on the augmented asset space can quite well approximate a dynamic strategy for medium-term horizons (up to five years), despite being much simpler to implement.

3.5 Investor Learning

In the portfolio choice literature, "learning" generally refers to the investor's gradually-better modeling of the generating distribution of asset returns, which may be conditional or not. In general, the optimal decision depends on the fact that we expect to learn about future changes in expected returns, which induces a negative hedging demand in the risky asset.[19] Kandel and Stambaugh (1996) and Barberis (2000) examine how asset return predictability and parameter estimation uncertainty affect the optimal allocations; both are found to induce sizable horizon effects, and bring substantial allocation differences that are exacerbated at long horizons. Xia (2001) discusses the effects of parameter uncertainty in a multiperiod context; it is found that the opportunity cost of ignoring predictability or learning is quite substantial. Brandt *et al.* (2005) propose a simulation approach to solve discrete-time dynamic portfolio choice problems involving non-standard preferences, a large number of assets and a large number of state variables, based on the well-known Longstaff and Schwartz (2001) approximation method originally proposed in the context of financial derivatives.

Finally, Skoulakis (2007) considers a fully Bayesian investor operating in discrete time and that solves a portfolio choice problem while simultaneously updating beliefs about the parameters of the generating distribution, considering that returns may (partially) be predictable. He finds that in the presence of predictability, learn-

---

[19] Put differently, this means that we desire less of the asset today, given that we expect to know more about its distribution with more observations in the future.

ing reduces, without completely eliminating, the positive hedging demands that are normally induced by predictability.

3.6 Common Extensions

Beyond the basic multiperiod framework of Samuelson and Merton, a large number of extensions have been proposed to address the shortcomings of the original formulations. In addition to the classical issues of consumption and labor income, extensive work has been pursued in the areas of non-standard preferences and utility functions, and characterization of the optimal policy in the presence of transaction costs, taxes and other frictions.

3.6.1 Intermediate Consumption and Labor Income

Intermediate consumption has traditionally been part of the multiperiod optimal investment problem since Samuelson (1969) and Merton (1969). In these settings, the problem is formulated so as to assume a single consumption good and postulates a time-separable utility over consumption. In continuous time, the investor's objective is then to jointly maximize the utility of the consumption path and terminal wealth,

$$\max_{\{C_t, \mathbf{w}_t\}_{t=0}^T} \mathbb{E}_0 \left[\int_0^T U_C(C_t, t)\, \mathrm{d}t + U_T(W_T) \right],$$

where $U_C(\cdot, \cdot)$ is the utility of the consumption rate C_t at time t, and U_T is the utility of the terminal wealth, subject to a modified budget constraint that accounts for consumption,

$$\mathrm{d}W_t = W_t \left(\mathbf{w}_t' \boldsymbol{\mu}_t^{\mathbf{P}} + r_f \right) \mathrm{d}t - C_t \mathrm{d}t + W_t \mathbf{w}_t' \mathbf{D}_t^{\mathbf{P}} \mathrm{d}\mathbf{B}_t^{\mathbf{P}}.$$

Non-stochastic labor income is just as easily incorporated by adding it into the budget constraint, as was shown in Merton (1971). The problem of stochastic labor income was treated by Koo (1998) and Viceira (2001) among others.

3.6.2 Non-Standard Preferences

Merton (1971) derives explicit solutions for the consumption–investment problem when investors have a time-separable utility over a consumption C that can be expressed as

$$U(C,t) = \exp(-\rho t) V(C),$$

with ρ a discount factor and V a utility function whose absolute risk aversion is positive and hyperbolic in its argument (i.e. belonging to the *hyperbolic absolute risk aversion*, or HARA, family),

$$V(C) = \frac{1-\gamma}{\gamma}\left(\frac{\beta C}{1-\gamma}+\eta\right)^{\gamma}, \tag{3.37}$$

subject to

$$\gamma \neq 1, \quad \beta > 0, \quad \frac{\beta C}{1-\gamma}+\eta > 0, \quad \eta = 1 \text{ if } \gamma = -\infty.$$

With suitable choice of parameters, the HARA family encompasses both Constant Absolute Risk Aversion (CARA) and Constant Relative Risk Aversion (CRRA) utilities.

As discussed in §2.4/p. 11, CRRA preferences are the only ones for which asset proportions are independent of wealth, and this—along with its analytical tractability—make it a popular choice in the literature. However, it can be shown (e.g. Campbell and Viceira 2002) that utility functions of this class intrinsically link the risk aversion with what is known as the *elasticity of intertemporal substitution* (the propensity to substitute consumption between periods). For this reason, Epstein and Zin (1989) introduced a class of recursive utility functions that generalize the CRRA class and admit independent risk aversion and coefficient of intertemporal substitution. Campbell and Viceira (1999) and Schroder and Skiadas (1999) analyze portfolio and consumption choices under this more general class of utility functions.

In a different vein, there has been in recent years an explosion of studies in the broad area of *behavioral finance*, where market participants are not assumed to always make rational choices.[20] In an asset allocation context, Shefrin and Statman (2000) apply the *prospect theory* of Kahneman and Tversky (1979) to construct a behavioral portfolio theory (BPT) and show that, in general, the "behavioral" efficient frontier does not coincide with the Markowitz one. In particular, BPT investors are simultaneously risk averse and risk seeking, and construct portfolios that consist of both bonds and lottery tickets. In recent work, Vlcek (2006) finds that in a two-period setting, an investor governed by prospect theory is not prone to the disposition effect,[21] his behavior instead essentially being driven by loss aversion: first-period gains cushion possible future losses and encourage increased risk-taking in the second period. Berkelaar *et al.* (2004) extend the martingale formulation of §3.4/p. 49 to analyze the optimal investment strategy of loss-averse investors. Brandt (2004) provides a broader survey of this literature.

---

[20] For comprehensive reviews of this vast field, see, e.g. Shefrin (2002) and Montier (2002); in an asset pricing context, Shefrin (2005) provides an in-depth treatment.

[21] The *disposition effect* refers to the empirical tendency of investors to prematurely sell winners and hold onto losers (Shefrin and Statman, 1985).

3.6.3 *Transaction Costs*

In continuous time, the absence of transaction costs causes the investor to continuously rebalance his portfolio, inducing an unrealistic level of trading activity. The effect of transaction costs in the continuous-time framework has long been studied, starting with Magill and Constantinides (1976) in the context of assets following a geometric Brownian motion with a HARA utility function in the presence of *proportional costs* (*cf.* eq. 2.12). They find that the optimal policy is characterized by an "envelope" around the time-t optimal portfolio weights (the targets). The optimal policy is not to trade when the current portfolio holdings are contained within the envelope, but to rebalance *up to the envelope* (but not up to the target) when they fall outside. This induces *random*, rather than continuous, portfolio rebalancing. This policy is found identical in functional form to the classical Bellman *et al.* (1955) ordering policy for the infinite-horizon multi-commodity inventory problem with proportional ordering costs. The intuitive justification for the presence of this envelope is that when existing holdings are close the the optimal (the targets at time t), there are only second-order utility gains to be made from adjusting the portfolio, but first-order transaction costs to bear.

Taksar *et al.* (1988) and Davis and Norman (1990) also studied related problems in the one-asset case, the latter relating it to the solution of a nonlinear free boundary problem. Cvitanić and Karatzas (1996) introduced a solution based on the martingale approach. Shreve and Soner (1994) analyzed the problem in terms of the viscosity solutions to the Hamilton-Jacobi-Bellman (HJB) equation. Leland (2000) generalizes the study to the multi-asset case (and also simultaneously considers capital-gain taxes), in the context of a *portfolio implementation* problem faced by a practitioner in which the target weights are provided exogenously. He characterizes an approximate no-trade region in terms of the 2^N corner points of the boundary surface. For a long time, it remained a practical problem that the runtime of the best existing solution methods to the HJB equation would grow super-exponentially with dimension N (the number of assets in the portfolio), making them impractical for portfolios of more than about $N = 4$ assets. Recently, Muthuraman and Zha (2008) proposed a simulation-based approach to tackle this problem which scales polynomially in dimension, while providing close fits to existing solutions.

The problem of *fixed transaction costs* was addressed by Eastham and Hastings (1988) in an optimal consumption–investment context; they derive a solution through quasi-variational inequalities of the value function. A numerical solution method was proposed by Atkinson *et al.* (1997) that scales well to moderate-sized portfolios (30 assets). Liu (2004) considered the case of a constant absolute risk aversion (CARA) investor dealing with both fixed and proportional transaction costs; his analysis reveals that costs can reduce the significance of asset return predictability on optimal portfolio rules.

Finally Morton and Pliska (1995) considered trading costs that are proportional to a fixed fraction of portfolio value, purely as a means to discourage frequent trading; they relate the solution to that of a stopping time problem. However, this "proportional-to-wealth" cost structure seems quite disconnected from that faced by real investors.

3.6.4 Taxes and Other Frictions

Capital gain taxes add another dimension to the transaction-cost issues. In many jurisdictions, including the United States, capital gains made when an asset is sold are taxable, although the tax rate may depend on the holding period of the asset (making short-term holdings more subject to penalty), and have a basis for calculating the tax amount that depends on the price at which the asset was originally bought (the tax basis). Furthermore, assets sold at a loss may offset gains from other assets. In contrast to simple handling of transaction costs, which depend only on local information, capital gain taxes complicate the solution of the backward dynamic programming equations (3.5) in the multiperiod case, since the buying price of an asset is unknown when "solving back in time". This can be overcome, approximately, by increasing the state space (recording, for each asset, not only the amount held in the portfolio, but also the original buy-date and buy-price), although this approach is clearly limited by the curse of dimensionality; moreover, an exact solution formulation grows exponentially with the number of time periods.[22] For single-period problems, Elton and Gruber (1978) first considered the question of capital gain taxes. In a multiperiod setting, Dammon *et al.* (2001; 2004) and Gallmeyer *et al.* (2006) approximate the tax basis by the weighted average purchase price. DeMiguel and Uppal (2005) show how to use the exact tax basis using a nonlinear programming formulation, and report that the certainty-equivalent loss from using an approximate basis is small. More recently, Osorio *et al.* (2008) apply a multistage stochastic programming approach (§4.6/p. 66) to this problem.

Cvitanić (2001) reviews the substantial literature that applies the Martingale formulation (§3.4/p. 49) to problems with frictions.

---

[22] Which requires recording the buy-date and buy-price for every transaction.

Chapter 4
Direct and Alternative Methods for Portfolio Choice

> *An economist is an expert who will know tomorrow why the things he predicted yesterday didn't happen today.*
>
> — *Laurence J. Peter*

I N THE SPIRIT of the original Markowitz methodology, all the portfolio allocation methods covered up to this point follow a functional separation that can be
summarized as

1. Estimate the (conditional) distribution (or moments thereof) of asset returns,
   from historical data and conditioning variables;
2. Construct an optimal portfolio (policy) by maximizing a utility function.

However, from a complete-system viewpoint, nothing prevents a more direct link
between conditioning variables and allocation decisions to be made. This can
be motivated from several perspectives. First, we already discussed (§2.8/p. 29) the
impact of estimation error on the stability of the resulting allocations, as well as
the complex arsenal of methods that have been proposed to remedy one aspect or
another of the problem. It can be shown that estimation errors compound in a multiperiod setting, making a bad problem worse (Brandt, 2004). Second, arguments
from statistical learning theory (Vapnik, 1998) can be made to the effect that to
solve problem X, given a limited amount of data (here, historical realizations of
financial series), one should not first attempt to solve a harder problem Y. In the
context of portfolio allocation, the really hard problem is the high-dimensional estimation of the conditional distribution of asset returns (which involves at least $O(N^k)$
quantities, where N is the number of assets, and k is the number of moments in the
distribution that we wish to represent), whereas the asset allocation itself involves
only $O(N)$ quantities—the weight to be given to each asset.

The idea of directly obtaining portfolio weights from explanatory variables has
first been explored in the machine learning community and has more recently
been studied in the financial economics literature as well. We also review "non-
allocation" approaches—mostly based on reinforcement learning—that do not at-

tempt to produce a genuine allocation of capital among assets but rather output a "long–short" decision aimed at short-term trading. Finally we conclude with an overview of approaches based on stochastic programming, studied mostly in operations research.

4.1 Machine Learning Approaches

The first applications of supervised learning algorithms to financial decision-making, and portfolio construction, problems have mainly focused on non-linear approaches to forecasting (e.g. Weigend and Gershenfeld 1993).[1] It goes without saying that these are simply generalizations of linear factor models (§2.7.1/p. 23) and do not sidestep the intrinsically difficult task of estimating the conditional distribution of asset returns.

Starting in the mid-1990's several asset-allocation approaches based on direct maximization of financial criteria started to appear. Choey and Weigend (1997) used a feedforward neural network (Rumelhart *et al.*, 1986) trained to directly maximize a Sharpe Ratio criterion to make an allocation decision between one risky (the DAX index of German stocks) and one riskless asset. Bengio (1997) trains a neural network on a profit criterion that accounts for transaction costs using a differentiable representation in the objective function, and compares against a network trained to optimize a forecasting criterion (the mean-squared error) on a basket of Canadian stocks; he reports significantly better out-of-sample risk-adjusted trading performance in favor of the financial criterion. In related work, Ghosn and Bengio (1997) analyze the parameter-sharing ability of multi-task learning to help improve forecasting performance across a universe of stocks, where each stock is viewed as a single task.

Chapados (2000) (see also Chapados and Bengio 2001) applied recurrent neural networks to a mean-VaR framework (*cf.* §2.5.3/p. 15), showing how the network can be trained to directly maximize expected return while satisfying a target portfolio risk constraint and minimize transaction costs. He compared against standard benchmarks including mean–variance optimization (where expected returns are forecast with a feedforward neural network and the covariance matrix is obtained by a standard RiskMetrics (1996) estimator) and obtains statistically significant out-of-sample financial performance in excess of the benchmark index when allocating to the 14 sub-sectors of the Canadian TSE-300 index.

Dunis *et al.* (2006a; 2006b) report good out-of-sample performance results using recurrent neural networks applied to trading commodity spread portfolios, beating standard feedforward neural networks and other benchmarks.

Zimmerman and colleagues have worked for a number of years on multi-tiered recurrent neural architectures that attempt to capture specific dynamical features

---

[1] More recent work include books by Shadbolt and Taylor (2002), Dunis *et al.* (2003), and McNelis (2005).

of financial time series. Zimmermann *et al.* (2001) apply a non-linear generalization of an $\text{ARMA}(1,1)$ model, termed an error-correcting neural network (ECNN), to forecast a set of price series. These forecasts are then subject to a second-level parametrized allocation function that determines an allocation $\mathbf{w}_{t,i}$ from a "softmax" transformation on asset-weighted excess returns. Let $\mathbf{f}_{t,i}$ be the expected-return forecast produced at time t for asset i by the ECNN; the portfolio weight is given by the second-level allocation function as

$$\mathbf{w}_{t,i} = \frac{\exp \mathbf{e}_{t,i}}{\sum_{j=1}^{N} \exp \mathbf{e}_{t,j}},$$

where $\mathbf{e}_{t,i}$ is the weighted average "excess" return of asset i against all other assets

$$\mathbf{e}_{t,i} = \sum_{j=1}^{N} \beta_{i,j}(\mathbf{f}_{t,i} - \mathbf{f}_{t,j}).$$

The parameters $\beta_{i,j}$ are optimized to maximize the in-sample total return subject to a deviation constraint from a benchmark. The authors claim that assets giving unreliable forecasts have their associated β coefficients pushed to zero, thereby implicitly controlling portfolio risk. They report risk-adjusted excess return on a stock–bond allocation task among the G7 countries with respect to an (unspecified) benchmark. More recently, the same group generalized these networks to operate at multiple time scales and applied them to the forecasting of foreign exchange (Zimmermann *et al.*, 2006a,b).

4.2 Parametric Portfolio Policies

Brandt *et al.* (2007) introduce an approach where the portfolio weight given to a stock directly depends on the *specific features* that characterize a stock though a parametrized functional form, $w_{i,t} = f(\mathbf{x}_{i,t}; \theta)$. In particular, they consider linear policies of the form

$$w_{i,t} = \bar{w}_{i,t} + \frac{1}{N_t} \theta' \hat{\mathbf{x}}_{i,t},$$

where $w_{i,t}$ is the weight of asset i at time t in the portfolio, $\bar{w}_{i,t}$ is a benchmark weight (e.g. $1/N$ or the weight in a capitalization-weighted market portfolio), θ is a fixed vector of coefficients (to be estimated) and $\hat{\mathbf{x}}_{i,t}$ is a vector of stock-dependent characteristics standardized cross-sectionally (at time t) to have a zero-mean and unit-standard deviation across all stocks. By construction (due to the standardization), the portfolio weights sum to one if the benchmark weights sum to one: the "correction term" $\frac{1}{N_t} \theta' \hat{\mathbf{x}}_{i,t}$ can be interpreted as a direct specification of the *active risk* of the position in asset i. The coefficients are optimized to maximize the expected utility of the one-period portfolio returns

$$\max_{\theta} \mathbb{E}_t \left[U(R_{P,t+1}) \right] = \max_{\theta} \mathbb{E}_t \left[U \left(\sum_{i=1}^{N_t} f(\mathbf{x}_{i,t}; \theta) \mathbf{R}_{i,t} \right) \right],$$

where the expectation is evaluated empirically on past data. Note that the parameters θ are fixed across both time and stocks. Note also that the approach applies effortlessly to a variable number of stocks during each period (indicated by the upper summation index N_t). Using only three conditioning variables[2] and all stocks from the CRSP–Compustat database from 1964 to 2002, this simple method generates statistically significant out-of-sample returns in excess of the benchmark, after transaction costs.

4.3 Nonparametric Portfolio Weights

Tying in more directly with the optimal multiperiod portfolio choice formulation, Brandt (1999) considers the sample analogues of the first-order optimality conditions[3] given by eq. (3.6). For single-period portfolio choice, this equation can be written

$$\mathbb{E}_t \left[U'(\mathbf{w}_t' \mathbf{R}_{t+1}) \mathbf{R}_{t+1} \right] = 0.$$

The idea is to estimate this expectation by a sample analogue (historical data), use a nonparametric estimator (Härdle, 1990) to weigh each observation according to how "close" it is to a given test state variable $\mathbf{z}$ and numerically solve for the portfolio weights $\mathbf{w}_t$ that satisfy the equation,

$$\hat{\mathbf{w}}_t(\mathbf{z}) = \left\{ \mathbf{w} : \frac{1}{T} \sum_{t=1}^{T} k_{h_T} (\mathbf{z}_t - \mathbf{z}) U'(\mathbf{w}' \mathbf{R}_{t+1}) \mathbf{R}_{t+1} = 0 \right\},$$

where $k_{h_T}(\cdot)$ is a kernel function (which we assumed is normalized), h_T is a kernel bandwidth parameter. The approach can be generalized to the multiperiod case by backward induction, assuming a CRRA utility function.

Unfortunately, this approach suffers from the curse of dimensionality in the number of state variables. Aït-Sahalia and Brandt (2001) propose an approach wherein an optimal linear projection down to a single state variable is found before applying the above kernel regression. This can be used to perform variable selection at the level of the state variables.

---

[2] Consisting of (i) the log market equity, (ii) the log book-to-market ratio, and (iii) the lagged one-year return, defined as the compounded return between months $t - 13$ and $t - 2$. The first two variables are used six months after their nominal validity date to ensure an adequate delay for the diffusion of financial statement information. Some experiments also added the slope of US interest rates yield curve as a conditioning variable for the other three.

[3] Also called Euler equations.

4.4 "Non-Allocation" Approaches

We call "non-allocation" approaches those that do not aim at solving a full portfolio problem (including taking advantage of diversification across assets) but perhaps the simpler problem of deciding whether the investor should be "long" (buy) or "short" (sell) in an asset. Most of the following approaches are based on reinforcement learning and approximate dynamic programming (Bertsekas and Tsitsiklis, 1996; Sutton and Barto, 1998; Si *et al.*, 2004; Powell, 2007).[4]

Neuneier (1996) uses a Q-learning algorithm (Watkins and Dayan, 1992) to learn the value function of a risk-neutral investor on the foreign-exchange and the stock market; Neuneier (1998) and Neuneier and Mihatsch (1999) generalize the approach to a multi-task setting and can derive multiperiod portfolio policies that account for transaction costs and risk-averse utility functions.

Ormoneit and Glynn (2001) introduce a non-parametric estimator of the value function for reinforcement learning and apply it to an allocation task between a risky and risk-free asset under logarithmic utility, where the decision is discretized (the fraction invested in the risky asset can be in the set $\{0, 0.1, 0.2, \ldots, 1.0\}$) and the state variable is proportional to the estimated risky asset volatility.

Moody *et al.* (1998) and Moody and Saffell (2001) introduce a "direct reinforcement" approach that sidesteps learning the value function and directly learns an allocation policy. They suggest an approximation scheme based on a Taylor series expansion to optimize a Sharpe Ratio criterion, which normally does not lend itself well to a reinforcement learning objective since it is not time-separable. More recently, Hens and Wöhrmann (2007) revisited the method in the context of strategic asset allocation between stocks and bonds for the US, UK, Germany, and Japan markets, for a power utility investor. The policy function is strictly determined by the forecasted yield spread between stocks and bonds (determined from average historical returns), which constitutes the only input variable. The learned policy suggests that this spread has significant explanatory power for market timing.

4.5 Information-Theoretic Approaches

Approaches based on information theory (Cover and Thomas, 2006) have also been investigated, although not traditionally by the financial economics community. Cover (1991) introduced *universal portfolios* which guarantee asymptotic performance equal to the best (in hindsight) *constant* portfolio weights.[5] Consider a fixed portfolio $\mathbf{w}, \sum_{i=1}^{N} \mathbf{w}_i = 1$, and let $S_k(\mathbf{w})$ be the cumulative portfolio return over a fixed horizon $t = 1 \ldots k$,

---

[4] Note that this survey must omit coverage of the vast fields of "automated trading systems". See, e.g. Kaufman (1998), for an introduction.

[5] Note that the strategy of keeping constant portfolio weights implies continuous rebalancing of the portfolio to keep the actual portfolio weights equal to their (constant) targets: as prices change, so do portfolio weights, which implies the necessity of rebalancing.

$$S_k(\mathbf{w}) \triangleq \prod_{t=1}^{k} \mathbf{w}'(1 + \mathbf{R}_t).$$

Let S_T^* be the maximum achievable wealth over a horizon-T given price sequence,

$$S_T^* = \max_{\mathbf{w} \in \mathcal{W}} S_T(\mathbf{w})$$

where $\mathcal{W}$ consists of the set of nonnegative-weight portfolios whose weights sum to one,

$$\mathcal{W} = \left\{ \mathbf{w} \in \mathbb{R}^N : \mathbf{w}_i \geq 0, \sum_{i=1}^{N} \mathbf{w}_i = 1 \right\}.$$

The UNIVERSAL portfolio strategy is simply defined as the performance-weighted constant portfolio average over a past history, namely

$$\hat{\mathbf{w}}_1 = \left(\frac{1}{N}, \frac{1}{N}, \cdots, \frac{1}{N} \right) \qquad \text{and} \qquad \hat{\mathbf{w}}_{k+1} = \frac{\int_{\mathcal{W}} \mathbf{w}\, S_k(\mathbf{w})\, d\mathbf{w}}{\int_{\mathcal{W}} S_k(\mathbf{w})\, d\mathbf{w}}.$$

Denote by $\hat{S}_T$ the wealth achieved by the UNIVERSAL portfolio strategy over the horizon T. Cover proved that for arbitrary bounded price sequences, the wealth achieved by the UNIVERSAL strategy grows as that of the best constant portfolio weights,[6]

$$(1/n) \ln \hat{S}_T - (1/n) \ln S_T^* \to 0.$$

Remarkably, this result does not depend on any statistical assumption on the behavior of the price sequences. Cover and Ordentlich (1996) considered the addition of side information (i.e. explanatory variables, albeit discrete ones) and obtains precise bounds on the ratio of the wealth given by the universal portfolio to the best wealth achievable by a constant rebalanced portfolio given hindsight. Ordentlich and Cover (1998) extended the results to an adversarial setting with bounds on achievable wealth in a game wherein a participant must announce a causal portfolio strategy at the outset and an opponent is allowed to choose any stock market sequence and the best constant rebalanced portfolio for that sequence.

Blum and Kalai (1998) addressed the original lack of consideration of transaction costs in Cover's formulation. In subsequent work, they also presented an efficient randomized approximation of the original algorithm that overcomes its exponential-time complexity (Kalai and Vempala, 2002).

4.6 Stochastic Programming Approaches

Stochastic programming[7](Dantzig, 1955; Birge and Louveaux, 1997) is a generalization of mathematical programming to optimization problems involving random

---

[6] Which is only known in hindsight and therefore unachievable.

[7] Not to be confused with dynamic programming or stochastic dynamic programming.

variables. A basic formulation of the problem is the following *two-stage stochastic linear program with fixed recourse,*

$$\min_{\mathbf{x}} \quad \mathbf{c}'\mathbf{x} + \mathbb{E}_{\xi}\left[Q(\mathbf{x},\xi)\right] \tag{4.1}$$

$$\text{subject to} \quad \mathbf{A}\mathbf{x} = \mathbf{b},$$

$$\mathbf{x} \geq 0,$$

where

$$Q(\mathbf{x},\xi) = \min_{\mathbf{y}} \mathbf{q}'\mathbf{y}$$

$$\text{subject to} \quad \mathbf{W}\mathbf{y} = \mathbf{h} - \mathbf{T}\mathbf{x},$$

$$\mathbf{y} \geq 0$$

and $\xi' \triangleq (\mathbf{q}',\mathbf{h}',\text{vec}(\mathbf{T}))$ and the matrix $\mathbf{W}$ is assumed fixed. This problem is interpreted as follows:

- The play unfolds in two acts, separated by the disclosure of a random variable ξ.[8]
- In the first stage (**decision-making**), the decision-maker must choose variables $\mathbf{x}$ to minimize a cost function made up of two parts: an immediate linear cost $\mathbf{c}'\mathbf{x}$ and an expected future cost $Q(\mathbf{x},\xi)$ (that is only known during the second stage).

- In the second stage (**recourse**), the decision-maker has been revealed the random variable ξ and must act to minimize the consequences (second-stage cost function Q) of this state of affairs.

Hence, in the first stage, the decision-maker acts ahead of time *knowing that he will act optimally in the second stage* to make do as well as possible given the scenario that just occurred. If the space of the random variable ξ is discrete (finite number of scenarios), then the stochastic program (4.1) can be converted to a classical (albeit large) deterministic linear program.

Extensions of the problem (4.1) to multiple decision stages are of course possible. To illustrate the application of the approach to asset allocation, we introduce a simple example inspired by Birge and Louveaux (1997). We assume an allocation tasks between N assets, $i = 1,\ldots,N$ over $t = 1,\ldots,T$ discrete periods. During each period t, a *scenario* s_t may occur, which is defined by the realization of random returns for all assets. More specifically, let $\xi(i,t,s_1,\ldots,s_t)$ be the random return for asset i during period t for scenario s_t, which may also depend on all previous realizations $s_1,\ldots,s_{t-1}$. This yields a (non-recombining) scenario tree illustrated in Fig. 4.1 (right). Furthermore, complete scenarios are given a probability $p(s_1,\ldots,p_T)$, which is used in the objective function (see below).

We assume that the investor is governed by the piecewise linear concave utility function shown in Fig. 4.1 (left). This function can be interpreted as follows: at

---

[8] The realization of this random variable is traditionally called a *scenario* in this context.

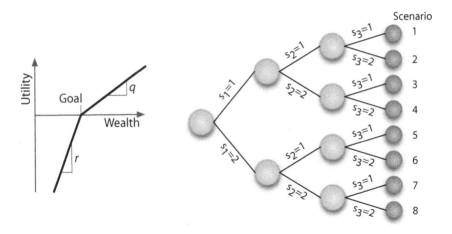

Fig. 4.1 Left: Terminal utility function for the stochastic programming asset allocation example; at the end of the investment horizon, if the wealth goal is reached, subsequent investment can be made for a return of $q\%$ (e.g. in the risk-free rate), otherwise money must be borrowed at a rate of $r\%$. **Right:** Scenario tree driving the model; at each period t, the joint stock–bond return is given by the scenario s_t. There is one decision node per period (large green circles), but decisions may depend on the entire realized history so far (i.e. the tree does not recombine). The smaller right-hand terminal nodes represent the final scenario outcomes.

horizon T the investor seeks to meet a financial goal G (for example, paying for Junior's college tuition). If this goal is met, the excess money can be invested at a yield of $q\%$, but if not, the missing money must be borrowed at a rate of $r\%$. Initially, the investor is endowed with W_0 dollars.

At the start of period t, the investor must make an allocation decision for each asset i, which is denoted $x(i,t,s_1,\dots,s_{t-1})$, and represents the dollar amount invested in asset i for the duration of the period. The "identity" of the decision variables depend on the scenario history until that point, and corresponds to the larger nodes in the scenario tree of Fig. 4.1.

The objective function is the value of the utility function realized for each complete scenario, weighted by the probability of that scenario; since the utility is piecewise-linear, it is split out into two terms by means of "surplus variables" w and y corresponding, respectively, to borrowing at a rate of $r\%$ and investing at a yield of $q\%$ (the surplus variables are defined as function of terminal wealth through constraints, below),

$$\max \sum_{s_T} \cdots \sum_{s_1} p(s_1,\dots,s_T)(-rw(s_1,\dots,s_T) + qy(s_1,\dots,s_T)).$$

The scenario probabilities $p(s_1,\dots,s_T)$ are specified by the modeler. The constraint for the first period is to invest the totality of initial wealth,

$$\sum_i x(i,1) = W_0.$$

The middle-period constraints, for $t = 2,\ldots,T-1$, are budget balance constraints: the wealth invested during period t must be that resulting from the investment during period $t-1$ (accrued by the yields earned during that period), with no possible intermediate reinvestment,

$$\sum_i -\xi(i,t-1,s_1,\ldots,s_{t-1})x(i,t-1,s_1,\ldots,s_{t-2})$$
$$+ \sum_i x(i,t,s_1,\ldots,s_{t-1}) = 0, \qquad \forall s_1,\ldots,s_{t-1}. \quad (4.2)$$

The last-period constraints take the terminal wealth generated in each scenario and split it among the surplus variables y and w for each scenario, depending on whether the accumulated wealth in that scenario is above or below the goal

$$\sum_i -\xi(i,T,s_1,\ldots,s_T)x(i,T,s_1,\ldots,s_{T-1})$$
$$- y(s_1,\ldots,s_T) + w(s_1,\ldots,s_T) = G, \qquad \forall s_1,\ldots,s_T. \quad (4.3)$$

We also force wealth to be positive in each period, along with the two surplus variables y and w,

$$x(i,t,s_1,\ldots,s_{t-1}) \geq 0 \qquad \forall i,t,s_1,\ldots,s_{t-1},$$
$$y(s_1,\ldots,s_T) \geq 0 \qquad \forall s_1,\ldots,s_{t-1},$$
$$w(s_1,\ldots,s_T) \geq 0 \qquad \forall s_1,\ldots,s_{t-1}.$$

This completes the formulation of the multistage stochastic program for this (admittedly simplified) asset allocation example. As presented, the optimization problem can easily be transformed into a (deterministic) linear program, yet the method also handles path-dependent events such as transaction costs and taxes (since the scenario tree of Fig. 4.1 does not recombine).

Dantzig and Infanger (1993) discusses the solution of multiperiod portfolio problems in the stochastic programming framework, and present algorithms based on a Benders decomposition of the linear program and Monte Carlo importance sampling. A survey of stochastic programming approaches in finance is presented by Yu *et al.* (2003). The book edited by Zenios (1993) provides additional useful references.

Due to its ability to model the complex real-world dependencies, stochastic programming has been widely applied to the problem of *asset–liability management* where a portfolio does not only consist of investments (future incoming cash flows) but also liabilities (future outgoing cash flows; for instance faced by an insurer whose written policies represent liabilities to be paid in the future, and who has reserves to invest optimally). Dempster *et al.* (2003) provide an in-depth presentation of the theory of stochastic programming to this problem, and followed up with an application to the management of minimum guaranteed return funds (Dempster *et al.*, 2007). The books edited by Zenios and Ziemba (2006) and Dempster *et al.* (2008) contain recent material on this topic.

It is perhaps unfortunate that stochastic programming approaches to portfolio optimization have mostly been studied by the operations research community, and are relatively unknown to financial economists. One can argue that this may be attributable to two factors: first, it was only recently that solution algorithms and computational power have become sufficient to enable solution to large-scale problems; still, the technological hurdle to get even simple stochastic programming models working may remain prohibitive to some. Second, a traditional emphasis of the solution methods studied by financial economists has been on the characterization of compact optimal policies, with significant concern paid to analytical tractability. Stochastic programming solutions, on the other hand, are mostly numerical and do not necessarily convey as much insight into optimal behavior. Yet, there appears to be much opportunity to combine the potential of multiple approaches, for instance integrating the methodological maturity of single-period modeling (e.g. the expected-return and risk models of §2.7/p. 23) with the ability of stochastic programming to cleanly handle a large number of real-world investment constraints over multiple periods.

Appendix A
Mathematical Complements

> *Little experience is sufficient to show that the traditional machinery of statistical processes is wholly unsuited to the needs of practical research.*
> — *Sir Ronald A. Fisher (1925)*

A.1 Minimization of a Quadratic Form Under Linear Equality Constraints

In §2/p. 7, we are faced with the problem of minimizing a quadratic form subject to linear equality constraints,

$$\mathbf{w}^* = \arg\min_{\mathbf{w}} \frac{1}{2} \mathbf{w}' \Sigma \mathbf{w} \tag{A.1}$$

$$\text{subject to} \quad \mathbf{A}\mathbf{w} = \mathbf{b}, \tag{A.2}$$

where $\mathbf{w} \in \mathbb{R}^N, \Sigma \in \mathbb{R}^{N \times N}, \mathbf{A} \in \mathbb{R}^{M \times N}, \mathbf{b} \in \mathbb{R}^M$. Let $\lambda \in \mathbb{R}^M$ be a vector of Lagrange multipliers. We consider the Lagrangian function

$$\mathscr{L}(\mathbf{w}, \lambda) = \frac{1}{2} \mathbf{w}' \Sigma \mathbf{w} + \lambda'(\mathbf{A}\mathbf{w} - \mathbf{b}). \tag{A.3}$$

The first-order conditions for optimality are obtained by differentiating (A.3) with respect to each variable and setting the resulting functions to zero,

$$\frac{\partial \mathscr{L}}{\partial \mathbf{w}'} = \Sigma \mathbf{w} + \mathbf{A}'\lambda = 0, \tag{A.4}$$

$$\frac{\partial \mathscr{L}}{\partial \lambda'} = \mathbf{A}\mathbf{w} - \mathbf{b} \quad = 0. \tag{A.5}$$

This implies the following system of equation, which can be written as a partitioned matrix equation

$$\begin{pmatrix} \Sigma & \mathbf{A}' \\ \mathbf{A} & \mathbf{0} \end{pmatrix} \begin{pmatrix} \mathbf{w} \\ \lambda \end{pmatrix} = \begin{pmatrix} \mathbf{0} \\ \mathbf{b} \end{pmatrix}. \tag{A.6}$$

Assuming that the inverse of $\begin{pmatrix} \Sigma & \mathbf{A}' \\ \mathbf{A} & \mathbf{0} \end{pmatrix}$ exists, the solution is given by

$$\begin{pmatrix} \mathbf{w} \\ \lambda \end{pmatrix} = \begin{pmatrix} \Sigma & \mathbf{A}' \\ \mathbf{A} & \mathbf{0} \end{pmatrix}^{-1} \begin{pmatrix} \mathbf{0} \\ \mathbf{b} \end{pmatrix}. \tag{A.7}$$

The inverse of the partitioned matrix is obtained as (*cf.* Greene 2007)[1]

$$\begin{pmatrix} \Sigma & \mathbf{A}' \\ \mathbf{A} & \mathbf{0} \end{pmatrix}^{-1} = \begin{pmatrix} \Sigma^{-1}(\mathbf{I} + \mathbf{A}'\mathbf{F}\mathbf{A}\Sigma^{-1}) & -\Sigma^{-1}\mathbf{A}'\mathbf{F} \\ -\mathbf{F}\mathbf{A}\Sigma^{-1} & \mathbf{F} \end{pmatrix}, \tag{A.8}$$

with $\mathbf{F} = -(\mathbf{A}\Sigma^{-1}\mathbf{A}')^{-1}$. It can be shown that this inverse exists if Σ^{-1} exists and $\mathbf{A}$ is of full rank. Substituting in eq. (A.7), we have

$$\mathbf{w} = -\Sigma^{-1}\mathbf{A}'\mathbf{F}\mathbf{b} \tag{A.9}$$

$$= \Sigma^{-1}\mathbf{A}'(\mathbf{A}\Sigma^{-1}\mathbf{A}')^{-1}\mathbf{b}. \tag{A.10}$$

A.2 Deriving the Tangency Portfolio

Consider the minimum-variance formulation of the mean-variance portfolio optimization problem (*cf.* §2.1/p. 7),

$$\min_{\mathbf{w}} \quad \frac{1}{2}\mathbf{w}'\Sigma\mathbf{w} \tag{A.11}$$

$$\text{subject to} \quad \mathbf{w}'\mu = \rho, \tag{A.12}$$

$$\mathbf{w}'\iota = 1, \tag{A.13}$$

where μ and Σ are respectively the vector of expected asset returns and covariance matrix between asset returns (assumed to be nonsingular), and ρ is a target portfolio return that remains unspecified. Let $\mathcal{W}_e$ be the *efficient frontier*, the set of all portfolios solving this problem (obtained by varying ρ). The tangency portfolio is the portfolio belonging to $\mathcal{W}_e$ having the largest return per unit of standard deviation,

$$\mathbf{w}^* = \arg\max_{\mathbf{w} \in \mathcal{W}_e} \frac{\mathbf{w}'\mu}{\sqrt{\mathbf{w}'\Sigma\mathbf{w}}}. \tag{A.14}$$

---

[1] This can be verified by direct multiplication, i.e. $\begin{pmatrix} \Sigma & \mathbf{A}' \\ \mathbf{A} & \mathbf{0} \end{pmatrix}^{-1} \begin{pmatrix} \Sigma & \mathbf{A}' \\ \mathbf{A} & \mathbf{0} \end{pmatrix} = \mathbf{I}.$

This portfolio is also seen as having a *maximal Sharpe Ratio* (Sharpe 1966, 1994).[2] The tangency portfolio is obtained in two steps: (i) characterization of the efficient frontier, and (ii) maximization of eq. (A.14).

A.2.1 Efficient Frontier

Portfolios on the efficient frontier are those that satisfy the following first-order conditions for optimality, obtained by incorporating constraints (A.12) and (A.13) into the objective through Lagrange multipliers and differentiating with respect to **w**,

$$\Sigma \mathbf{w} - \lambda_1 \mu - \lambda_2 \iota = 0,$$

yielding an optimal portfolio of the form

$$\mathbf{w}^* = \Sigma^{-1}(\lambda_1 \mu + \lambda_2 \iota). \tag{A.15}$$

The Lagrange multipliers are obtained by substituting this solution back into the constraints, and must jointly satisfy

$$\lambda_1 \mu' \Sigma^{-1} \mu + \lambda_2 \mu' \Sigma^{-1} \iota = \rho \tag{A.16}$$

and

$$\lambda_1 \iota' \Sigma^{-1} \mu + \lambda_2 \iota' \Sigma^{-1} \iota = 1. \tag{A.17}$$

A.2.2 Maximization of the Sharpe Ratio

The first-order conditions for the solution of eq. (A.14) are obtained by differentiating with respect to **w**, yielding

$$\frac{\mu}{\sqrt{\mathbf{w}' \Sigma \mathbf{w}}} - \frac{\mathbf{w}' \mu}{(\mathbf{w}' \Sigma \mathbf{w})^{3/2}} \Sigma \mathbf{w} = 0,$$

which simplifies to

---

[2] It should be noted that absent the sum-to-one constraint, pure maximization of the Sharpe Ratio is an ill-posed problem. To see this, consider scaling the positions of portfolio P by a positive constant γ. This yields Sharpe Ratio

$$SR_{\gamma P} = \frac{\mathbb{E}[\gamma R_P]}{\sqrt{\mathrm{Var}[\gamma R_P]}} = \frac{\gamma \mathbb{E}[R_P]}{\gamma \sqrt{\mathrm{Var}[R_P]}} = SR_P.$$

Hence in order to maximize the Sharpe Ratio, it is necessary to choose the scaling factor, which corresponds to establishing the target portfolio risk level. Alternatively, enforcing a sum-to-one constraint specifies a risk level as well. Sharpe Ratio maximization, despite occasional claims to the contrary, does not absolve one from specifying risk preferences.

$$(\mathbf{w}'\mu)\Sigma\mathbf{w} = (\mathbf{w}'\Sigma\mathbf{w})\mu$$

and implies

$$\mathbf{w}^* = \frac{\mathbf{w}^{*\prime}\Sigma\mathbf{w}^*}{\mathbf{w}^{*\prime}\mu}\Sigma^{-1}\mu. \tag{A.18}$$

Since the tangency portfolio belongs to the efficient frontier, it must have the general form given by eq. (A.15). Comparing with eq. (A.18), this is only possible if $\lambda_2 = 0$. Substituting into eq. (A.17), we obtain

$$\lambda_1 = \frac{1}{\iota'\Sigma^{-1}\mu},$$

which finally yields the desired **functional form for the tangency portfolio**,

$$\mathbf{w}^* = \frac{\Sigma^{-1}\mu}{\iota'\Sigma^{-1}\mu}.$$

This can be interpreted as follows: the numerator assigns weight to "virtual assets" (formed by decorrelated linear combinations of the original assets) proportionally of their individual expected-return/variance ratio, and transforms them back into the space of original assets. The denominator acts as a normalization term (sum of the elements) to ensure that the elements of $\mathbf{w}^*$ sum to one.

If there is a risk-free asset, the same result obtains if we instead consider μ to be the vector of expected *excess* asset returns. This amounts to shifting down the efficient frontier by a constant equal to the risk-free rate.

A.3 Itô's Lemma

Itô's celebrated lemma (Itô, 1951) is central to any study of continuous-time financial models involving stochastic differential equations. It shows how to express the differential of a (sufficiently smooth) function of a random process. Standard textbooks on stochastic calculus cover this material, such as Shreve (2005a, 2005b). Its use in the context of continuous-time portfolio optimization appears in §3.2/p. 45.

A.3.1 Wiener Processes

Define the stochastic process $Z(t)$ as

$$Z(t+h) = Z(t) + y(t)\sqrt{h},$$

where $y(t)$ is a process of IID standard normal variables (i.e. zero-mean and unit variance) and $h > 0$. It can be observed that $Z(t+h) \sim N(z(t),h)$. In the limit as $h \to 0$, the increment $Z(t+h) - Z(t)$ follows a standard Wiener process and is defined as

$$dZ \triangleq y(t)\sqrt{dt}.$$

The Wiener process serves as a building block of more complex processes in continuous-time finance. For example, a simple process modeling the evolution of stock prices (sometimes termed a *geometric Brownian motion*) is

$$\frac{dP_t}{P_t} = \mu(t)dt + \sigma(t)dZ.$$

The interpretation of this equation is that the *relative* change in the price P_t of the stock is given by the sum of a deterministic return $\mu(t)$ and a stochastic return proportional to $\sigma(t)$. Both $\mu(t)$ and $\sigma(t)$ are here deterministic functions of time. The process P_t is continuous but nowhere differentiable.

A.3.2 One-Dimensional Case

Consider the Markov random process $X(t)$ specified as

$$dX(t) = \mu(X(t),t)dt + \sigma(X(t),t)dZ(t),$$

where $dZ(t)$ is a standard Wiener process, denoted

$$dZ(t) = u(t)\sqrt{dt}, \qquad u(t) \overset{\text{iid}}{\sim} N(0,1).$$

The functions $\mu(X,t)$ and $\sigma(X,t)$ are, respectively, deterministic functions giving the *drift rate* and *volatility* of the process.

Lemma 1. *Let $f : \mathbb{R} \times [0,\infty] \mapsto \mathbb{R}$ a square-integrable function. Then the random process $f(X(t),t)$ is given by the differential*

$$df = \frac{\partial f(X,t)}{\partial X}dX + \frac{\partial f(X,t)}{\partial t}dt + \frac{1}{2}\frac{\partial^2 f(X,t)}{\partial X^2}(dX)^2 \qquad (A.19)$$

where the product of differentials is given by the multiplication rules

$$(dZ)^2 = dt, \qquad dZ\,dt = 0, \qquad (dt)^2 = 0.$$

A.3.3 Multi-Dimensional Case

The generalization to the multi-dimensional case obtains readily. We shall consider the case where the dimensionality of the process $\mathbf{X}(t)$ is the same as the underlying sources of uncertainty. Let $\mathbf{X}(t) \in \mathbb{R}^N$ be specified as

$$dX(t) = \mu(X(t),t)dt + \sigma(X,t)dZ(t),$$

where $dZ(t)$ is a correlated Wiener process

$$dZ(t) = u(t)\sqrt{dt}, \qquad u(t) \overset{iid}{\sim} N(0,\Gamma)$$

with $(\Gamma)_{i,j} \equiv \rho_{i,j}$ is the *correlation* between variables Z_i and Z_j; we assume them to have unit variance. The functions μ and σ are suitably generalized to be N-vector-valued.

Lemma 2. *Let $f : \mathbb{R}^N \times [0,\infty] \mapsto \mathbb{R}$ a square-integrable function. Then the random process $f(X(t),t)$ is given by the differential*

$$df = \sum_{i=1}^{N} \frac{\partial f(X,t)}{\partial X_i} dX_i + \frac{\partial f(X,t)}{\partial t} dt + \frac{1}{2} \sum_{i=1}^{N} \sum_{j=1}^{N} \frac{\partial^2 f(X,t)}{\partial X_i \partial X_j} dX_i dX_j \qquad (A.20)$$

where the product of differentials is given by the multiplication rules

$$(dZ_i dZ_j) = \rho_{i,j} dt, \qquad dZ_i dt = 0, \qquad (dt)^2 = 0.$$

Glossary

APT   Arbitrage Pricing Theory. (Page 24)

CAPM   Capital Asset Pricing Model. (Page 23)

CARA   Constant Absolute Risk Aversion. (Page 57)

CML   Capital Market Line. (Page 11)

CRRA   Constant Relative Risk Aversion. (Page 57)

DRAWDOWN Worst decline suffered by an investment from its peak value.(Page 4)

EFFICIENCY A portfolio **w** is said to be *efficient* if it is the lowest-variance portfolio for a given level of expected return. (Page 8)

HARA   Hyperbolic Absolute Risk Aversion. (Page 48)

IID   Independent and Identically Distributed. (Page 41)

LONG POSITION The buying of a security such as a stock, commodity or currency, with the expectation that the asset will rise in value. Opposite of *short position*. (Page 17)

MPT   Modern Portfolio Theory. (Page 2)

SHORT POSITION The sale of a borrowed security such as a stock, commodity or currency with the expectation that the asset will fall in value. Opposite of *long position*. (Page 17)

SIMPLE RETURN The *simple rate of return* of an asset during period t is given by $R_t = \frac{P_t}{P_{t-1}} - 1$ where P_t is the price of the asset at time t. (Page 5)

VAR   Vector Autoregressive. (Page 26)

References

Alexander, G. J. and Baptista, A. M. (2002). Economic implications of using a mean-var model for portfolio selection. *Journal of Economic Dynamics & Control*, **26**, 1159–1193.

Arrow, K. J. (1965). *Aspects of the Theory of Risk-Bearing*. Yrjö Jahnsson Lectures. Yrjö Jahnsson Foundation, Helsinki, Finland.

Artzner, P., Delbaen, F., Eber, J.-M., and Heath, D. (1999). Coherent measures of risk. *Mathematical Finance*, **3**, 203–228.

Atkinson, C., Pliska, S. R., and Wilmott, P. (1997). Portfolio management with transaction costs. *Proceedings of the Royal Society: Mathematical, Physical and Engineering Sciences*, **453**(1958), 551–562.

Aït-Sahalia, Y. and Brandt, M. W. (2001). Variable selection for portfolio choice. *Journal of Finance*, **56**(4), 1297–1351.

Aït-Sahalia, Y. and Brandt, M. W. (2007). Consumption and portfolio choice with option-implied state prices. Working Paper, Princeton University.

Banz, R. W. (1981). The relationship between return and market value of common stocks. *Journal of Financial Economics*, **9**(1), 3–18.

Barberis, N. (2000). Investing for the long run when returns are predictable. *Journal of Finance*, **55**(1), 225–264.

Barra (1998). *Risk Model Handbook, United States Equity: Version 3*. MSCI Barra, Berkeley, CA.

Basu, S. (1983). The relationship between earnings yield, market value, and return for nyse common stocks: Further evidence. *Journal of Financial Economics*, **12**, 129–156.

Bellman, R. E. (1957). *Dynamic Programming*. Princeton University Press, Princeton, NJ.

Bellman, R. E., Glicksberg, I., and Gross, O. (1955). On the optimal inventory equation. *Management Science*, **2**, 83–104.

Ben-Tal, A. and Nemirovski, A. (1999). Robust solutions of uncertain linear programs. *Operations Research Letters*, **25**, 1–13.

Bengio, Y. (1997). Using a financial training criterion rather than a prediction criterion. *International Journal of Neural Systems*, **8**(4), 433–443.

Berkelaar, A. B., Kouwenberg, R., and Post, T. (2004). Optimal portfolio choice under loss aversion. *Review of Economics and Statistics*, **86**(4), 973–987.

Bernoulli, D. (1738). Specimen theoriæ novæ de mensura sortis. In *Commentarii Academiæ Scientiarum Imperialis Petropolitannæ*. Translated from Latin into English by L. Sommer, "Exposition of a New Theory on the Measurement of Risk," Econometrica 22, 23–36.

Bertsekas, D. P. (2000). *Nonlinear Programming*. Athena Scientific, Belmont, MA, second edition.

Bertsekas, D. P. (2005). *Dynamic Programming and Optimal Control*, volume I. Athena Scientific, Belmont, MA, third edition.

Bertsekas, D. P. and Tsitsiklis, J. N. (1996). *Neuro-Dynamic Programming*. Athena Scientific, Belmont, MA.

Bertsimas, D. and Pachamanova, D. A. (2008). Robust multiperiod portfolio management in the presence of transaction costs. *Computers and Operations Research*, **35**(1), 3–17.

Best, M. J. and Grauer, R. (1991). On the sensitivity of mean-variance efficient portfolios to changes in asset means: Some analytical and computational results. *Review of Financial Studies*, **4**, 315–342.

Bevan, A. and Winkelmann, K. (1998). Using the black-litterman global asset allocation model: Three years of practical experience. Technical report, Fixed Income Research, Goldman Sachs.

Birge, J. R. and Louveaux, F. (1997). *Introduction to Stochastic Programming*. Springer, New York, NY.

Bishop, C. M. (2006). *Pattern Recognition and Machine Learning*. Springer, New York, NY.

Black, F. and Litterman, R. (1992). Global portfolio optimization. *Financial Analysts Journal*, **48**(5), 28–43.

Blum, A. and Kalai, A. (1998). Universal portfolios with and without transaction costs. *Machine Learning*, **30**(1), 23–30.

Bodie, Z., Kane, A., and Marcus, A. J. (2004). *Investments*. Irwin McGraw-Hill, Boston, MA, sixth edition.

Boyd, S. and Vandenberghe, L. (2004). *Convex Optimization*. Cambridge University Press, Cambridge, UK.

Brandt, M. W. (1999). Estimating portfolio and consumption choice: A conditional euler equations approach. *Journal of Finance*, **54**(5), 1609–1645.

Brandt, M. W. (2004). Portfolio choice problems. Technical report, Duke University, Fuqua School of Business. Working Paper.

Brandt, M. W. and Santa-Clara, P. (2006). Dynamic portfolio selection by augmenting the asset space. *Journal of Finance*, **61**(5), 2187–2217.

Brandt, M. W., Goyal, A., Santa-Clara, P., and Stroud, J. R. (2005). A simulation approach to dynamic portfolio choice with an application to learning about return predictability. *Review of Financial Studies*, **18**(3), 831–871.

Brandt, M. W., Santa-Clara, P., and Valkanov, R. (2007). Parametric portfolio policies: Exploiting characteristics in the cross section of equity returns. Working Paper, Duke University.

Brennan, M. J., Schwartz, E. S., and Lagnado, R. (1997). Strategic asset allocation. *Journal of Economic Dynamics and Control*, **21**, 1377–1403.

Brown, S. J. (1978). The portfolio choice problem: Comparison of certainty equivalent and optimal bayes portfolios. *Communications in Statistics: Simulation and Computation*, **B7**, 321–334.

Campbell, J. Y. (1991). A variance decomposition for stock returns. *Economic Journal*, **101**(405), 157–179.

Campbell, J. Y. and Shiller, R. J. (1988). The dividend price ratio and expectations of future dividends and discount factors. *Review of Financial Studies*, **1**(3), 195–228.

Campbell, J. Y. and Viceira, L. M. (1999). Consumption and portfolio decisions when expected returns are time varying. *Quarterly Journal of Economics*, **114**(2), 433–495.

Campbell, J. Y. and Viceira, L. M. (2002). *Strategic Asset Allocation: Portfolio Choice for Long-Term Investors*. Oxford University Press, New York, NY.

Campbell, J. Y., Lo, A. W., and MacKinlay, A. C. (1997). *The Econometrics of Financial Markets*. Princeton University Press, Princeton, NJ, USA.

Carhart, M. M. (1997). On persistence in mutual fund performance. *Journal of Finance*, **52**(1), 57–82.

Chang, T.-J., Meade, N., Beasley, J. E., and Sharaiha, Y. M. (2000). Heuristics for cardinality constrained portfolio optimization. *Computers and Operations Research*, **27**, 1271–1302.

Chapados, N. (2000). *Critères d'optimisation d'algorithmes d'apprentissage en gestion de portefeuille*. Master's thesis, Département d'informatique et de recherche opérationnelle, Université de Montréal, Montréal, Canada.

Chapados, N. (2010). *Sequential Machine Learning Approaches for Portfolio Management*. Ph.D. thesis, University of Montreal, Department of Computer Science and Operations Research. Available at http://hdl.handle.net/1866/3578.

Chapados, N. and Bengio, Y. (2001). Cost functions and model combination for VaR-based asset allocation using neural networks. *IEEE Transactions on Neural Networks*, **12**(4), 890–906.

Choey, M. and Weigend, A. S. (1997). Nonlinear trading models through sharpe ratio maximization. In A. S. Weigend, Y. Abu-Mostafa, and A.-P. Refenes, editors, *Decision Technologies for Financial Engineering: Proceedings of the Fourth International Conference on Neural Networks in the Capital Markets (NNCM '96)*, pages 3–22. World Scientific Publishing.

Chopra, V. K. and Ziemba, W. T. (1993). The effect of errors in means, variances, and covariances on optimal portfolio choice. *Journal of Portfolio Management*, **19**(2), 6–11.

Chow, G. and Kritzman, M. (2002). Value at risk for portfolios with short positions. *Journal of Portfolio Management*, **???**, 73–81.

Claus, J. and Thomas, J. (2001). Equity premia as low as three percent? evidence from analysts' earnings forecasts for domestic and international stock markets. *Journal of Finance*, **56**(5), 1629–1666.

Consigli, G. (2004). Estimation of tail risk and portfolio optimisation with respect to extreme measures. In G. Szegö, editor, *Risk Measures for the 21st Century*, page Chapter 18, Chichester. John Wiley & Sons.

Cootner, P. H., editor (1964). *The Random Character of Stock Market Prices*. MIT Press, Cambridge, MA.

Cover, T. M. (1991). Universal portfolios. *Mathematical Finance*, **1**(1), 1–29.

Cover, T. M. and Ordentlich, E. (1996). Universal portfolios with side information. *IEEE Transactions on Information Theory*, **42**(2), 348–363.

Cover, T. M. and Thomas, J. A. (2006). *Elements of Information Theory*. John Wiley & Sons, New York, NY, second edition.

Cox, J. C. and Huang, C.-F. (1989). Optimum consumption and portfolio policies when asset prices follow a diffusion process. *Journal of Economic Theory*, **49**, 33–83.

Cox, J. C. and Huang, C.-F. (1991). A variational problem arising in financial economics. *Journal of Mathematical Economics*, **20**, 465–487.

Crichfield, T., Dyckman, T., and Lakonishok, J. (1978). An evaluation of security analysts' forecasts. *The Accounting Review*, **53**(3), 651–668.

Cvitanić, J. (2001). Theory of portfolio optimization in markets with frictions. In E. Jouini, J. Cvitanić, and M. Musiela, editors, *Handbooks in Mathematical Finance: Option Pricing, Interest Rates and Risk Management*, pages 577–631, Cambridge, UK. Cambridge University Press.

Cvitanić, J. and Karatzas, I. (1996). Hedging and portfolio optimization under transaction costs: A martingale approach. *Mathematical Finance*, **6**(2), 133–165.

Cvitanić, J., Goukasian, L., and Zapatero, F. (2003). Monte carlo computation of optimal portfolios in complete markets. *Journal of Economic Dynamics and Control*, **27**(6), 971–986.

Cvitanić, J., Lazrak, A., and Wang, T. (2008). Implications of the sharpe ratio as a performance measure in multi-period settings. *Journal of Economic Dynamics and Control*, **32**(5), 1622–1649.

Dammon, R. M., Spatt, C. S., and Zhang, H. H. (2001). Optimal consumption and investment with capital gains taxes. *Review of Financial Studies*, **14**(3), 583–616.

Dantzig, G. B. (1955). Linear programming under uncertainty. *Management Science*, **1**(3–4), 197–206.

Dantzig, G. B. and Infanger, G. (1993). Multi-stage stochastic linear programs for portfolio optimization. *Annals of Operations Research*, **45**(1-4), 59–76.

Davis, M. H. A. and Norman, A. R. (1990). Portfolio selection with transaction costs. *Mathematics of Operations Research*, **15**(4), 676–713.

De Bondt, W. F. M. and Thaler, R. (1985). Does the stock market overreact? *Journal of Finance*, **40**(3), 793–805.

DeMiguel, V. and Uppal, R. (2005). Portfolio investment with the exact tax basis via nonlinear programming. *Management Science*, **51**(2), 277–290.

DeMiguel, V., Garlappi, L., and Uppal, R. (2009). Optimal versus naive diversification: How inefficient is the 1/n portfolio strategy? *Review of Financial Studies*, **22**(5), 1915–1953.

Dempster, M. A. H., Germano, M., Medova, E. A., and Villaverde, M. (2003). Global asset liability management. *British Actuarial Journal*, **9**(1), 137–216.

Dempster, M. A. H., Germano, M., Medova, E. A., Rietbergen, M. I., Sandrini, F., and Scrowston, M. (2007). Designing minimum guaranteed return funds. *Quantitative Finance*, **7**(2), 245–256.

Dempster, M. A. H., Pflug, G., and Mitra, G., editors (2008). *Quantitative Fund Management*. Financial Mathematics Series. Chapman & Hall / CRC.

Detemple, J. B., Garcia, R., and Rindisbacher, M. (2003). A monte carlo method for optimal portfolios. *Journal of Finance*, **58**(1), 401–446.

Duffie, D. (2001). *Dynamic Asset Pricing Theory*. Princeton University Press, Princeton, NJ, third edition.

Dunis, C. L., Laws, J., and Naïm, P., editors (2003). *Applied Quantitative Methods for Trading and Investment*. John Wiley & Sons, Chichester, UK.

Dunis, C. L., Laws, J., and Evans, B. (2006a). Modelling and trading the soybean-oil crush spread with recurrent and higher order networks: A comparative analysis. *Neural Network World*, **16**, 193–213.

Dunis, C. L., Laws, J., and Evans, B. (2006b). Trading futures spread portfolios: Applications of higher order and recurrent networks. Technical report, Centre for International Banking, Economics and Finance; Liverpool John Moores University. www.cibef.com.

Eastham, J. F. and Hastings, K. J. (1988). Optimal impulse control of portfolios. *Mathematics of Operations Research*, **13**(4), 588–605.

Edwards, E. O. and Bell, P. W. (1961). *Theory and Measurement of Business Income*. University of California Press, Berkeley, CA.

Elton, E. J. and Gruber, M. J. (1978). Taxes and portfolio composition. *Journal of Financial Economics*, **6**, 399–410.

Epstein, L. G. and Zin, S. E. (1989). Substitution, risk aversion, and the temporal behavior of consumption and asset returns: A theoretical framework. *Econometrica*, **57**(4), 937–969.

Estrada, J. (2007). Mean-semivariance optimization: A heuristic approach. Technical report, IESE Business School, Spain.

Fabozzi, F. J., Focardi, S. M., and Kolm, P. N. (2006). *Financial Modeling of the Equity Market: From CAPM to Cointegration*. The Frank J. Fabozzi Series. John Wiley & Sons, Hoboken, NJ.

Fabozzi, F. J., Kolm, P. N., Pachamanova, D. A., and Focardi, S. M. (2007). *Robust Portfolio Optimization and Management*. The Frank J. Fabozzi Series. John Wiley & Sons, Hoboken, NJ.

Fama, E. F. (1970). Efficient capital markets: A review of theory and empirical work. *Journal of Finance*, **25**(2), 383–417. Papers and Proceedings of the Twenty-Eighth Annual Meeting of the American Finance Association New York, N.Y. December, 28–30, 1969.

Fama, E. F. and French, K. R. (1988). Dividend yields and expected stock returns. *Journal of Financial Economics*, **22**(1), 3–25.

Fama, E. F. and French, K. R. (1992). The cross-section of expected stock returns. *Journal of Finance*, **47**(2), 427–465.

Fama, E. F. and French, K. R. (1993). Common risk factors in the returns on stocks and bonds. *Journal of Financial Economics*, **33**, 3–56.

Fama, E. F. and French, K. R. (1995). Size and book-to-market factors in earnings and returns. *Journal of Finance*, **50**(1), 131–155.

Fama, E. F. and French, K. R. (1996). Multifactor explanations of asset pricing anomalies. *Journal of Finance*, **51**(1), 55–84.

Farinelli, S., Rossello, D., and Tobiletti, L. (2006). Computational asset allocation using one-sided and two-sided variability measures. In *Computational Science – ICCS 2006*, Lecture Notes in Computer Science 3994, pages 324–331, Berlin / Heidelberg. Springer.

Fisher, I. (1906). *The Nature of Capital and Income*. Macmillan, London.

Fisher, R. A. (1925). *Statistical Methods for Research Workers*. Oliver and Boyd, Edinburgh.

Friesen, G. and Weller, P. A. (2006). Quantifying cognitive biases in analyst earnings forecasts. *Journal of Financial Markets*, **9**(4), 333–365.

Frost, P. A. and Savarino, J. E. (1986). An empirical bayes approach to efficient portfolio selection. *Journal of Financial and Quantitative Analysis*, **21**, 293–305.

Frost, P. A. and Savarino, J. E. (1988). For better performance: Constrain portfolio weights. *Journal of Portfolio Management*, **15**, 29–34.

Gallmeyer, M. F., Kaniel, R., and Tompaidis, S. (2006). Tax management strategies with multiple risky assets. *Journal of Financial Economics*, **80**(2), 243–291.

Ghosn, J. and Bengio, Y. (1997). Multi-task learning for stock selection. In M. I. Jordan, M. C. Mozer, and T. Petsche, editors, *Advances in Neural Information Processing Systems 9*, pages 946–952. MIT Press, Cambridge, MA.

Givoly, D. and Lakonishok, J. (1984). The quality of analysts' forecasts of earnings. *Financial Analysts Journal*, **40**(5), 40–47.

Goldfarb, D. and Iyengar, G. (2003). Robust portfolio selection problems. *Mathematics of Operations Research*, **28**, 1–38.

Gordon, M. (1962). *The Investment, Financing, and Valuation of the Corporation*. Irwin Publishing, Homewood, IL.

Graham, B. and Dodd, D. L. (1934). *Security Analysis: Principles and Techniques*. McGraw-Hill, Columbus, OH.

Greene, W. H. (2007). *Econometric Analysis*. Prentice Hall, Englewood Cliffs, NJ, sixth edition.

Grinold, R. C. and Kahn, R. N. (2000). *Active Portfolio Management*. McGraw Hill.

Halldórsson, B. V. and Tütüncü, R. H. (2003). An interior-point method for a class of saddle-point problems. *Journal of Optimization Theory and Applications*, **116**(3), 559–590.

Hamilton, J. D. (1994). *Time Series Analysis*. Princeton University Press, Princeton, NJ.

Harrison, M. and Kreps, D. (1979). Martingales and arbitrage in multiperiod securities markets. *Journal of Economic Theory*, **20**, 381–408.

Hastie, T., Tibshirani, R., and Friedman, J. (2009). *Elements of Statistical Learning*. Springer, Berlin, New York, second edition.

Hens, T. and Wöhrmann, P. (2007). Strategic asset allocation and market timing: A reinforcement learning approach. *Computational Economics*, **29**, 369–381.

Härdle, W. (1990). *Applied Nonparametric Regression*. Cambridge University Press, New York, NY.

Itô, K. (1951). On stochastic differential equations. *Memoirs of the American Mathematical Sociecy*, **4**, 1–51.

Jagannathan, R. and Ma, T. (2003). Risk reduction in large portfolios: Why imposing the wrong constraints helps. *Journal of Finance*, **58**, 1651–1683.

James, W. and Stein, C. (1961). Estimation with quadratic loss. In *Proceedings of the Fourth Berkeley Symposium on Mathematics and Statistics*, pages 361–379.

Jegadeesh, N. and Titman, S. (1993). Returns to buying winners and selling losers: Implications for stock market efficiency. *Journal of Finance*, **48**(1), 65–91.

Jin, H., Markowitz, H. M., and Zhou, X. Y. (2006). A note on semivariance. *Mathematical Finance*, **16**(1), 53–61.

Jobson, J. D. and Korkie, B. M. (1980). Estimation of markowitz efficient portfolios. *Journal of the American Statistical Association*, **75**, 544–554.

Jobson, J. D. and Ratti, V. (1979). Improved estimation for markowitz portfolios using james-stein type estimators. In *Proceedings of the American Statistical Association, Business and Economic Statistics Section*, pages 279–284.

Jorion, P. (1986). Bayes-stein estimation for portfolio analysis. *Journal of Financial and Quantitative Analysis*, **21**, 279–292.

Jorion, P. (2003). Portfolio optimization with tracking-error constraints. *Financial Analysts Journal*, **59**, 70–82.

Kahneman, D. and Tversky, A. (1979). Prospect theory: An analysis of decision under risk. *Econometrica*, **47**(2), 263–292.

Kalai, A. and Vempala, S. (2002). Efficient algorithms for universal portfolios. *Journal of Machine Learning Research*, **3**, 423–440.

Kallberg, J. G. and Ziemba, W. T. (1983). Comparison of alternative utility functions in portfolio selection problems. *Management Science*, **29**(11), 1257–1276.

Kandel, S. and Stambaugh, R. F. (1996). On the predictability of stock returns: An asset-allocation perspective. *Journal of Finance*, **51**(2), 385–424.

Karatzas, I., Lehoczky, J. P., and Shreve, S. E. (1987). Optimal portfolio and consumption decisions for a "small investor" on a finite horizon. *SIAM Journal of Control and Optimization*, **25**(6), 1557–1586.

Kaufman, P. J. (1998). *Trading Systems and Methods*. John Wiley & Sons, third edition.

Kellerer, H., Mansini, R., and Speranza, M. G. (2000). Selecting portfolios with fixed costs and minimum transaction lots. *Annals of Operations Research*, **99**, 287–304.

Kim, T. S. and Omberg, E. (1996). Dynamic nonmyopic portfolio behavior. *The Review of Financial Studies*, **9**(1), 141–161.

Kissell, R. and Glantz, M. (2003). *Optimal Trading Strategies: Quantitative Approaches for Managing Market Impact and Trading Risk*. AMACOM/American Management Association, New York, NY.

Klein, R. and Bawa, V. (1976). The effect of estimation risk on optimal portfolio choice. *Journal of Financial Economics*, **3**, 215–231.

Koo, H. K. (1998). Consumption and portfolio selection with labor income: A continuous time approach. *Mathematical Finance*, **8**(1), 49–65.

Korn, R. and Trautmann, S. (1995). Continuous-time portfolio optimization under terminal wealth constraints. *ZOR – Mathematical Methods of Operations Research*, **42**(1), 69–92.

Krokhmal, P., Palmquist, J., and Uryasev, S. (2002). Portfolio optimization with conditional value-at-risk objective and constraints. *The Journal of Risk*, **4**(2), 11–27.

Lakonishok, J., Shleifer, A., and Vishny, R. W. (1994). Contrarian investment, extrapolation, and risk. *Journal of Finance*, **49**(5), 1541–1578.

Laloux, L., Cizeau, P., Bouchaud, J.-P., and Potters, M. (1999). Noise dressing of financial correlation matrices. *Physics Review Letter*, **83**, 1467–1470.

Ledoit, O. and Wolf, M. (2004). Honey, i shrunk the sample covariance matrix. *Journal of Portfolio Management*, **30**(4), 110–119.

Leippold, M., Trojani, F., and Vanini, P. (2004). A geometric approach to multiperiod mean variance optimization of assets and liabilities. *Journal of Economic Dynamics and Control*, **28**(6), 1079–1113.

Leland, H. E. (2000). Optimal portfolio implementation with transaction costs and capital gains taxes. Working Paper, University of California, Berkeley.

Levy, H. and Markowitz, H. M. (1979). Approximating expected utility by a function of mean and variance. *American Economic Review*, **69**(3), 308–317.

Li, D. and Ng, W.-L. (2000). Optimal dynamic portfolio selection: Multiperiod mean-variance formulation. *Mathematical Finance*, **10**(3), 387–406.

Lintner, J. (1965). The valuation of risk assets and the selection of risky investments in stock portfolios and capital budgets. *The Review of Economics and Statistics*, **47**(1), 13–37.

Litterman, R. M., editor (2003). *Modern Investment Management*. Wiley Finance.

Liu, H. (2004). Optimal consumption and investment with transaction costs and multiple risky assets. *Journal of Finance*, **59**(1), 289–338.

Lo, A. W. and MacKinlay, A. C. (1999). *A Non-Random Walk Down Wall Street*. Princeton University Press, Princeton, NJ.

Longstaff, F. A. and Schwartz, E. S. (2001). Valuing american options by simulation: A simple least-squares approach. *Review of Financial Studies*, **14**(1), 113–147.

Luenberger, D. G. and Ye, Y. (2007). *Linear and Nonlinear Programming*. International Series in Operations Research & Management Science. Springer, third edition.

Magill, M. J. P. and Constantinides, G. M. (1976). Portfolio selection with transactions costs. *Journal Of Economic Theory*, **13**(2), 245–263.

Malevergne, Y. and Sornette, D. (2005a). *Extreme Financial Risks: From Dependence to Risk Management*. Springer, Berlin / Heidelberg.

Malevergne, Y. and Sornette, D. (2005b). High-order moments and cumulants of multivariate weibull asset returns distributions: Analytical theory and empirical tests ii. *Finance Letters, Special Issue: Modeling of the Equity Market*, 3(1), 54–63.

Markowitz, H. M. (1952). Portfolio selection. *Journal of Finance*, 7, 77–91.

Markowitz, H. M. (1959). *Portfolio Selection: Efficient Diversification of Investment*. John Wiley & Sons, New York, London, Sydney.

Markowitz, H. M. (1999). The early history of portfolio theory: 1600–1960. *Financial Analysts Journal*, 55, 5–16.

Markowitz, H. M. and Usmen, N. (2003). Resampled frontiers versus diffuse bayes: An experiment. *Journal of Investment Management*, 1, 9–25.

McNelis, P. D. (2005). *Neural Networks in Finance: Gaining Predictive Edge in the Market*. Academic Press, Burlington, MA.

Merton, R. C. (1969). Lifetime portfolio selection under uncertainty: the continuous-time case. *Review of Economics and Statistics*, 51(3), 247–257. Reprinted in (Merton, 1990, Chapter 4).

Merton, R. C. (1971). Optimum consumption and portfolio rules in a continuous-time model. *Journal of Economic Theory*, 3, 373–413. Reprinted in (Merton, 1990, Chapter 5).

Merton, R. C. (1972). An analytic derivation of the efficient portfolio frontier. *The Journal of Financial and Quantitative Analysis*, 7(4), 1851–1872.

Merton, R. C. (1973). An intertemporal capital asset pricing model. *Econometrica*, 41(5), 867–887. Reprinted in (Merton, 1990, Chapter 15).

Merton, R. C. (1990). *Continuous-Time Finance*. Blackwell Publishers, Cambridge, MA.

Merton, R. C. and Samuelson, P. A. (1974). Fallacy of the log-normal approximation to optimal portfolio decision making over many periods. *Journal of Financial Economics*, 1, 67–94.

Michaud, R. O. (1989). The markowitz optimization enigma: Is optimized optimal? *Financial Analysts Journal*, 45, 31–42.

Michaud, R. O. (1998). *Efficient Asset Management: A Practical Guide to Stock Portfolio Optimization and Asset Allocation*. Oxford University Press, Oxford, UK.

Mittnik, S., Rachev, S. T., and Schwartz, E. S. (2003). Value at risk and asset allocation with stable return distributions. *Allgemeines Statistisches Archiv*, 86, 53–67.

Montier, J. (2002). *Behavioural Finance: Insights into Irrational Minds and Market*. John Wiley & Sons, Chichester, UK.

Moody, J. and Saffell, M. (2001). Learning to trade via direct reinforcement. *IEEE Transactions on Neural Networks*, 12(4), 875–889.

Moody, J., Wu, L., Liao, Y., and Saffell, M. (1998). Performance functions and reinforcement learning for trading systems and portfolios. *Journal of Forecasting*, 17, 441–470.

Morton, A. J. and Pliska, S. R. (1995). Optimal portfolio management with fixed transaction costs. *Mathematical Finance*, 5(4), 337–356.

Mossin, J. (1966). Equilibrium in a capital asset market. *Econometrica*, 34(4), 768–783.

Mossin, J. (1968). Optimal multiperiod portfolio policies. *Journal of Business*, 41(2), 215–229.

Muthuraman, K. and Zha, H. (2008). Simulation-based portfolio optimization for large portfolios with transaction costs. *Mathematical Finance*, 18(1), 115–134.

Nawrocki, D. (1999). A brief history of downside risk measures. *Journal of Investing*, 8, 9–25.

Neuneier, R. (1996). Optimal asset allocation using adaptive dynamic programming. In D. S. Touretzky, M. C. Mozer, and M. E. Hasselmo, editors, *Advances in Neural Information Processing Systems 8*, pages 952–958. MIT Press.

Neuneier, R. (1998). Enhancing q-learning for optimal asset allocation. In M. I. Jordan, M. J. Kearns, and S. A. Solla, editors, *Advances in Neural Information Processing Systems 10*, pages 936–942. The MIT Press.

Neuneier, R. and Mihatsch, O. (1999). Risk sensitive reinforcement learning. In M. J. Kearns, S. A. Solla, and D. A. Cohn, editors, *Advances in Neural Information Processing Systems 11*, pages 1031–1037. The MIT Press.

Newey, W. K. and West, K. D. (1987). A simple, positive semi-definite, heteroskedasticity and autocorrelation consistent covariance matrix. *Econometrica*, 55(3), 703–708.

Ocone, D. L. and Karatzas, I. (1991). A generalized clark representation formula, with application to optimal portfolios. *Stochastics and Stochastics Reports*, **34**(3), 187–220.

Ohlson, J. A. (1995). Earnings, book values, and dividends in equity valuation. *Contemporary Accounting Research*, **11**(2), 661–687.

Ordentlich, E. and Cover, T. M. (1998). The cost of achieving the best portfolio in hindsight. *Mathematics of Operations Research*, **23**(4), 960–982.

Ormoneit, D. and Glynn, P. W. (2001). Kernel-based reinforcement learning in average-cost problems: An application to optimal portfolio choice. In T. K. Leen, T. G. Dietterich, and V. Tresp, editors, *Advances in Neural Information Processing Systems 13*, pages 1068–1074. MIT Press.

Osorio, M. A., Gülpinar, N., and Rustem, B. (2008). A general framework for multistage mean-variance post-tax optimization. *Annals of Operations Research*, **157**(1), 3–23.

Pan, J. and Poteshman, A. M. (2006). The information in option volume for future stock prices. *Review of Financial Studies*, **19**(3), 871–908.

Phelps, E. S. (1962). The accumulation of risky capital: A sequential utility analysis. *Econometrica*, **30**(4), 729–743.

Philips, T. K. (2003). Estimating expected returns. *Journal of Investing*, pages 49–57.

Pliska, S. R. (1986). A stochastic calculus model of continuous trading: Optimal portfolios. *Mathematics of Operations Research*, **11**(2), 371–382.

Powell, W. B. (2007). *Approximate Dynamic Programming: Solving the Curses of Dimensionality*. John Wiley & Sons, Hoboken, NJ.

Pratt, J. W. (1964). Risk aversion in the small and in the large. *Econometrica*, **32**, 122–136.

Qian, E. E., Hua, R. H., and Sorensen, E. H. (2007). *Quantitative Equity Portfolio Management: Modern Techniques and Applications*. CRC Financial Mathematics Series. Chapman & Hall.

Rachev, S. T., Menn, C., and Fabozzi, F. J. (2005). *Fat-Tailed and Skewed Asset Return Distributions: Implications for Risk Management, Portfolio Selection, and Option Pricing*. John Wiley & Sons, Hoboken, NJ.

Rau-Bredow, H. (2004). Value-at-risk, expected shortfall and marginal risk contribution. In G. Szegö, editor, *Risk Measures for the 21st Century*, pages 61–68, Chichester. John Wiley & Sons.

RiskMetrics (1996). Riskmetrics—technical document. Technical report, J.P. Morgan, New York, NY. http://www.riskmetrics.com.

Robert M. Dammon, C. S. S. and Zhang, H. H. (2004). Optimal asset location and allocation with taxable and tax-deferred investing. *Journal of Finance*, **59**(3), 999–1037.

Rosenberg, B., Reid, K., and Lanstein, R. (1985). Persuasive evidence of market inefficiency. *Journal of Portfolio Management*, **11**, 9–17.

Ross, S. A. (1976). The arbitrage theory of capital asset pricing. *Journal of Economic Theory*, **13**(3), 341–360.

Roy, A. D. (1952). Safety-first and the holding of assets. *Econometrica*, **20**(3), 431–449.

Rubinstein, M. (2002). Markowitz's "portfolio selection": A fifty-year retrospective. *Journal of Finance*, **57**(3), 1041–1045.

Rumelhart, D. E., Hinton, G. E., and Williams, R. J. (1986). *Learning Internal Representations by Error Propagation*, volume Parallel Distributed Processing: Explorations in the Microstructure of Cognition, chapter 8, pages 310–362. MIT Press, Cambridge, MA.

Samuelson, P. A. (1965). Proof that properly anticipated prices fluctuate randomly. *Industrial Management Review*, **6**, 41–49.

Samuelson, P. A. (1969). Lifetime portfolio selection by dynamic stochastic programming. *Review of Economics and Statistics*, **51**(3), 239–246.

Satchell, S., editor (2007). *Forecasting Expected Returns in the Financial Markets*. Academic Press, London, UK.

Scherer, B. (2002). Portfolio resampling: Review and critique. *Financial Analysts Journal*, **58**, 98–109.

Schroder, M. and Skiadas, C. (1999). Optimal consumption and portfolio selection with stochastic differential utility. *Journal of Economic Theory*, **89**(1), 68–126.

Shadbolt, J. and Taylor, J. G. (2002). *Neural Networks and the Financial Markets: Predicting, Combining and Portfolio Optimisation*. Springer, Berlin, New York.

Sharpe, W. F. (1963). A simplified model for portfolio analysis. *Management Science*, **9**, 277–293.

Sharpe, W. F. (1964). Capital asset prices: A theory of market equilibrium under conditions of risk. *Journal of Finance*, **19**(3), 425–442.

Sharpe, W. F. (1966). Mutual fund performance. *Journal of Business*, **39**(1), 119–138.

Sharpe, W. F. (1994). The sharpe ratio. *The Journal of Portfolio Management*, **21**(1), 49–58.

Shefrin, H. (2002). *Beyond Greed and Fear: Understanding Behavioral Finance and the Psychology of Investing*. Oxford University Press, Oxford, UK.

Shefrin, H. (2005). *A Behavioral Approach to Asset Pricing*. Academic Press, Burlington, MA.

Shefrin, H. and Statman, M. (1985). The disposition to sell winners too early and ride losers too long: Theory and evidence. *Journal of Finance*, **40**(3), 777–790.

Shefrin, H. and Statman, M. (2000). Behavioral portfolio theory. *Journal of Financial and Quantitative Analysis*, **35**(2), 127–151.

Shreve, S. E. (2005a). *Stochastic Calculus for Finance I: The Binomial Asset Pricing Model*. Springer, New York, NY.

Shreve, S. E. (2005b). *Stochastic Calculus for Finance II: Continuous-Time Models*. Springer, New York, NY.

Shreve, S. E. and Soner, H. M. (1994). Optimal investment and consumption with transaction costs. *Annals of Applied Probability*, **4**(3), 609–692.

Si, J., Barto, A. G., Powell, W. B., and Wunsch, D., editors (2004). *Handbook of Learning and Approximate Dynamic Programming*. IEEE Press Series on Computational Intelligence. Wiley–IEEE Press.

Skoulakis, G. (2007). Dynamic portfolio choice with bayesian learning. Working Paper, University of Maryland.

Stein, C. (1956). Inadmissibility of the usual estimator for the mean of multivariate normal distribution. In *Prooceedings of the Third Berkeley Symposium on Mathematical Statistics and Probability*, pages 197–206.

Sutton, R. S. and Barto, A. G. (1998). *Reinforcement Learning: An Introduction*. MIT Press, Cambridge, MA.

Taksar, M., Klass, M. J., and Assaf, D. (1988). A diffusion model for optimal portfolio selection in the presence of brokerage fees. *Mathematics of Operations Research*, **13**(2), 277–294.

Theil, H. and Goldberger, A. (1961). On pure and mixed estimation in economics. *International Economic Review*, **2**, 65–78.

Tobin, J. (1958). Liquidity preference as a behavior towards risk. *Review of Economic Studies*, **67**, 65–86.

Tobin, J. (1965). The theory of portfolio selection. In F. H. Hahn and F. P. R. Brechling, editors, *The Theory of Interest Rates*, London. Macmillan.

Tütüncü, R. H. and Koenig, M. (November 2004). Robust asset allocation. *Annals of Operations Research*, **132**, 157–187.

Vapnik, V. N. (1998). *Statistical Learning Theory*. John Wiley & Sons, New York, London, Sydney.

Viceira, L. M. (2001). Optimal portfolio choice for long-horizon investors with nontradable labor income. *Journal of Finance*, **56**(2), 433–470.

Vlcek, M. (2006). Portfolio choice with loss aversion, asymmetric risk-taking behavior and segregation of riskless opportunities. Swiss Finance Institute Research Paper No. 27. Available at http://ssrn.com/abstract=947078.

Wachter, J. A. (2002). Portfolio and consumption decisions under mean-reverting returns: An exact solution for complete markets. *Journal of Financial and Quantitative Analysis*, **37**(1), 63–91.

Wagner, W. H. and Edwards, M. (1998). Implementing investment strategies: The art and science of investing. In F. J. Fabozzi, editor, *Active Equity Portfolio Management*, page 186, Hoboken, NJ. John Wiley & Sons.

Watkins, C. J. C. H. and Dayan, P. (1992). *q*-learning. *Machine Learning*, **8**, 279–292.

Weigend, A. S. and Gershenfeld, N. (1993). *Time Series Prediction: Forecasting the future and understanding the past*. Addison-Wesley, Reading, MA, USA.

Williams, J. B. (1938). *The Theory of Investment Value*. Harvard University Press, Cambridge, MA.

Wilmott, P. (2006). *Paul Wilmott on Quantitative Finance*. John Wiley & Sons, second edition.

Wolsey, L. A. and Nemhauser, G. L. (1999). *Integer and Combinatorial Optimization*. Wiley-Interscience, New York, NY.

Xia, Y. (2001). Learning about predictability: The effects of parameter uncertainty on dynamic asset allocation. *Journal of Finance*, **56**(1), 205–246.

Yu, L.-Y., Ji, X.-D., and Wang, S.-Y. (2003). Stochastic programming models in financial optimization: A survey. *Advanced Modeling and Optimization*, **5**(1), 1–26.

Zellner, Z. A. and Chetty, V. K. (1965). Prediction and decision problems in regression models from the bayesian point of view. *Journal of the American Statistical Association*, **60**, 608–615.

Zenios, S. A., editor (1993). *Financial Optimization*. Cambridge University Press, Cambridge, UK.

Zenios, S. A. and Ziemba, W. T., editors (2006). *Handbook of Asset and Liability Management, Volume 1: Theory and Methodology*. Handbooks in Finance. North Holland, Amsterdam, Netherlands.

Zhou, X. Y. (2000). Continuous-time mean-variance portfolio selection: A stochastic lq framework. *Applied Mathematics and Optimization*, **42**(1), 19–33.

Zimmermann, H.-G., Neuneier, R., and Grothmann, R. (2001). Active portfolio-management based on error correction neural networks. In T. G. Dietterich, S. Becker, and Z. Ghahramani, editors, *Advances in Neural Information Processing Systems 14*, pages 1465–1472. MIT Press.

Zimmermann, H.-G., Grothmann, R., Schäfer, A. M., and Tietz, C. (2006a). Modeling large dynamical systems with dynamical consistent neural networks. In S. Haykin, J. C. Príncipe, T. J. Sejnowski, and J. McWhirter, editors, *New Directions in Statistical Signal Processing: From Systems to Brains*, pages 203–242, Cambridge, MA. MIT Press.

Zimmermann, H.-G., Bertolini, L., Grothmann, R., Schäfer, A. M., and Tietz, C. (2006b). A technical trading indicator based on dynamical consistent neural networks. In *ICANN (2)*, pages 654–663.

Author Index

Subject Index